CLIMATE CHANGE
SOLUTION
365 Ways to Fight Climate Change and
a Roadmap to Net Zero India by 2050

CLIMATE CHANGE SOLUTION

365 Ways to Fight Climate Change and
a Roadmap to Net Zero India by 2050

Himanshu Nayak

BLACK EAGLE BOOKS
Dublin, USA | Bhubaneswar, India

Black Eagle Books
USA address:
7464 Wisdom Lane
Dublin, OH 43016

India address:
E/312, Trident Galaxy, Kalinga Nagar,
Bhubaneswar-751003, Odisha, India

E-mail: info@blackeaglebooks.org
Website: www.blackeaglebooks.org

First International Edition Published by
Black Eagle Books, 2025

CLIMATE CHANGE SOLUTION
365 Ways to Fight Climate Change and a Roadmap to Net Zero India by 2050
by **Himanshu Nayak**

Cover & Interior Design: Ezy's Publication

ISBN- 978-1-64560-688-8 (Paperback)

Printed in the United States of America

This book is dedicated to my elder son, Rudraditya—a truly special soul—and to my loving wife, whose unwavering love and support made this journey possible. It was written in moments borrowed from our time together, and it belongs as much to them as it does to me. Thank you for the sacrifices, the patience, and the unconditional love that made it all possible.

Preface

Climate change is the defining challenge of our time. Its consequences are far-reaching, affecting every aspect of life—our environment, economy, health, and future. While scientific advancements, industrialization, and urbanization have propelled humanity forward, they have also led to unprecedented environmental degradation. Rising global temperatures, extreme weather events, and shifting ecosystems are stark reminders of the urgent need for action.

This book is not just an exploration of the problem but a roadmap to solutions. It is a call to action for individuals, businesses, and policymakers to take meaningful steps towards sustainability. The COVID-19 pandemic demonstrated that humans can adapt rapidly when faced with an immediate crisis. Pollution levels dropped, nature thrived, and people embraced a simpler, less consumption-driven lifestyle. These lessons provide a crucial blueprint for addressing climate change.

As one of the fastest-growing economies, India has a pivotal role in global climate action. With a commitment to achieving net-zero emissions by 2070, the nation has set ambitious targets to transition toward a low-carbon economy. However, this goal can only be met through collective effort and widespread public participation.

Each of us has the power to contribute by reducing our carbon footprint, adopting renewable energy, and actively restoring our environment.

This book presents a comprehensive analysis of climate change, its causes and impacts, and the necessary strategies to mitigate its effects. It aims to inspire readers to take proactive measures, reinforcing the idea that change starts with us. By working together, we can combat climate change and secure a sustainable, thriving future for generations to come.

Let this book be a guide, a source of motivation, and a reminder that the time to act is now.

CONTENTS

Chapter 1

Overview

Introduction

"The world has enough for everyone's need, but not enough for everyone's greed." Mahatma Gandhi

There are approximately 8.7 million species of living organisms on Earth, with humans being the most intelligent and capable. Born free, humans possess remarkable cognitive abilities, emotions, and an understanding of their surroundings, enabling them to rise above other species and dominate the planet.

However, the rapid acceleration of scientific and technological advancements, population growth, urbanization, industrialization, and the adoption of luxurious lifestyles, as well as the use of chemical fertilizers in agriculture and deforestation, have led to severe environmental degradation. These activities have contributed to global warming and climate change, with far-reaching consequences.

While the effects of climate change may not always be immediately visible, they are becoming increasingly evident. We are witnessing more frequent heatwaves, rising surface temperatures, prolonged droughts, irregular rainfall patterns, devastating floods, intense cyclones and hurricanes, melting glaciers and snow, rising sea levels, beach erosion, ocean acidification, ecosystem shifts, health crises, water scarcity, food supply challenges, and a rise in vector-borne diseases. These impacts are causing significant economic losses and, beyond a certain temperature threshold, could become irreversible, leading to the collapse of ecological, social, and financial systems. It is, therefore, essential to limit global temperature rise to 1.5°C and transition to a low-carbon economy in the coming decades.

Climate change poses a severe economic threat to India, potentially causing a GDP loss of 3% to 10% by 2100. Under a business-as-usual scenario, India could face a GDP per capita loss of 2.6% by 2030, 6.7% by 2050, and 16.9% by 2100.

Unlike the immediate and visible impacts of the COVID-19 pandemic, the effects of climate change are more gradual but far more destructive and deadly. While COVID-19 is a tangible threat with symptoms like fever, cough, and breathing difficulties, it is largely recoverable within 7-14 days. In contrast, pollution-related premature deaths accumulate over time, with long-lasting consequences. The urgency of our response to the pandemic highlights the need for similar, if not greater, action to address the climate crisis.

During the COVID-19 lockdown, pollution levels dropped dramatically. People could see the Himalayas from afar, animals roamed freely, and humans were confined to their homes—much like the animals they often cage. This period demonstrated that humans can adapt and sacrifice for survival. It proved that a low-carbon lifestyle is achievable, as people consumed less, stayed home, and avoided unnecessary travel. This experience provides a blueprint for combating climate change.

If Indians can tolerate a 30-40-day lockdown, prioritizing health and well-being over consumption and profit, they can adopt a net-zero carbon lifestyle. The pandemic showed that people act swiftly when the threat to life is immediate and undeniable.

Climate change presents a far more catastrophic and dangerous challenge than COVID-19. India, a rapidly developing nation, is projected to see its population and GDP grow significantly by 2070, with its GDP expected to rise from USD 3 trillion in 2023 to approximately USD 27 trillion by 2070. While India's per capita energy consumption is only one-third of the global average, its energy demand is set to increase substantially with economic growth.

India is committed to combating climate change and has set an ambitious target to achieve net-zero emissions by 2070. To support this goal, the country has outlined the following targets for 2030:

- **Non-Fossil Fuel Capacity:** Increase the share of non-fossil fuel capacity to 50% of its total installed electricity capacity and 500GW
- **Non-Fossil Energy Capacity:** 500 GW of non-fossil energy capacity.
- **Reduce emission intensity:** A 45% reduction in carbon intensity relative to 2005 levels and a reduction of carbon emissions by 1 billion tons.
- **Carbon Sink:** Cumulative 2.5-3 GtCO2 carbon sink through additional Tree plantation

The time has come for us to act as we did during the COVID-19 pandemic. The solutions to climate change lie in the same behaviors and commitments that helped us tackle the pandemic.

Previously, India aimed to achieve net-zero emissions by 2050. However, due to slow progress and limited public participation, this target has been extended to 2070. Without accelerated efforts, this deadline may be further delayed.

What, then, is the solution?

The solution is simple: Take action in our daily lives as we did during COVID-19. Every individual and business must strive to become net-zero emission-compliant. The following steps are essential:

1. Reduce your carbon footprint in everyday lifestyle.
2. Transition to and produce energy from renewable sources.
3. Plant more trees to absorb carbon dioxide from the atmosphere.

By adopting these measures and leading the way toward sustainability, we can mitigate climate change, protect our environment, and ensure a safer, healthier future for future generations.

This comprehensive overview explores climate change, its causes, effects, challenges, and potential solutions at the individual, business, government, and international levels. It aims to inspire everyone to take corrective actions in their own lives, contributing to the goal of achieving Net Zero India by 2050. By doing so, India can lead the global effort to mitigate climate change, much like the collective response seen during the COVID-19 pandemic.

Mind-set and Ignorance: Key Drivers of Climate Change
"We cannot solve our problems with the same thinking we used when we created them." Albert Einstein

With rapid advancements in science and technology, humans are increasingly focusing on short-term solutions for immediate happiness, often overlooking the long-term consequences. This mind-set leads to significant harm over time. For example, someone might avoid going to the doctor for a routine check-up or delay treating a minor health issue because it doesn't cause immediate pain or discomfort. They might think, "It's not urgent; I'll deal with it later." However, by ignoring it, minor health problems can escalate into severe, chronic conditions that could have been prevented with early intervention. This short-term avoidance ultimately creates long-term consequences for both health and well-being.

This mindset is also evident in how we address environmental issues. Urbanization and population growth have led to deforestation, pollution, and habitat loss to meet immediate needs like infrastructure and food. Similarly, luxury lifestyles provide instant gratification but contribute to future environmental crises.

Rising greenhouse gas emissions in India exacerbate climate change, and its effects are becoming increasingly visible. While ignoring climate change offers short-term benefits—such as avoiding lifestyle changes, keeping production costs low, and maintaining outdated energy systems—the long-term consequences are devastating.

The impacts of climate change are undeniable: rising temperatures, frequent heatwaves, unpredictable rainfall patterns, melting glaciers, air pollution, and worsening health conditions, particularly cardiovascular diseases. Economic losses are mounting due to extreme heat, yet many believe governments alone will handle the situation. This mindset must change.

To address climate change effectively, we need a collective mindset shift. Relying solely on government action is insufficient. Awareness, education, and responsibility at individual, business, government, and international levels are crucial to mitigating the crisis and securing a sustainable future.

Greenhouse Gases: Key Contributors to Climate Change
"The truth is, the natural world is changing. And we are totally dependent on that world. It provides our food, water, and air. It is the most precious thing we have, and it is our responsibility to care for it."
David Attenborough, renowned naturalist and broadcaster

Greenhouse gases absorb and emit heat energy from sunlight, contributing to the greenhouse effect and global warming. The primary greenhouse gases are:

Since the Industrial Revolution, which began around 1750, human activities have significantly increased the concentration of CO_2 and other greenhouse gases in the atmosphere, thereby exacerbating climate change.

According to data from Climate Watch, a platform by the World Resources Institute is a partner of the UNFCCC and the World Bank Group.

Annual CO_2 emission for Top 10 Countries of World since 1990 to 2021:

Name of Country	Annual CO_2 Emission (MtCO2e) on 1990	Annual CO_2 Emission (MtCO2e) on 2021	Changes from 1990 to 2021	% of Changes from 1990 to 2021
China	1826.08	10849.18	9023.1	494.12
USA	4435.58	4385.23	-50.35	-1.14
India	**340.83**	**2395.19**	**2054.36**	**602.75**
Russia	1811.01	1211.29	-599.72	-33.12
Indonesia	752.03	1060.49	308.46	41.02
Brazil	1236.44	849.86	-386.58	-31.27
Japan	1015.11	990.74	-24.37	-2.40

Iran	179.3	712.56	533.26	297.41
Germany	922.56	626.16	-296.4	-32.13
Canada	501.54	568.64	67.1	13.38

India's greenhouse gas emissions have increased by 233.44% from 1990 to 2021, while CO_2 emissions have surged by 602.75% during the same period. In 2021, India emitted 3,419.89 metric tonnes of greenhouse gases and 2,395.19 metric tonnes of carbon dioxide equivalent (tCO_2e). Despite these figures, India's emission rate remains relatively high.

China is the world's largest CO_2 emitter, releasing 10,849.18 tCO_2e in 2021, up from 1,826.08 tCO_2e in 1990—a 494.12% increase over the past 31 years. The United States is the second-largest CO_2 emitter, with 4,385.23 tCO_2e in 2021, compared to 4,435.58 tCO_2e in 1990.

In India, the per capita availability of Greenhouse Gas Equivalent (tCO_2e) was 2.43 metric tonnes in 2021, compared to 1.18 $MtCO_2e$ in 1990. Similarly, the per capita Carbon Dioxide Equivalent (CO_2e) was 1.70 metric tonnes in 2021, up from 0.39 $MtCO_2e$ in 1990. This represents an increase of 105.93% in per capita GHG emissions and 335.90% in CO_2 emissions over the past 31 years, despite significant population growth. Current trends indicate a continued rise in emissions.

Greenhouse gas emissions from the top four emitting countries account for over 50% of global emissions: China (26%), the United States (13%), the European Union (7.8%), and India (6.7%).

In 2016, India's total greenhouse gas emissions were 2,839 $MtCO_2e$ excluding Land Use, Land-Use Change, and Forestry (LULUCF), and 2,531 $MtCO_2e$ when including LULUCF.

Per GDP CO_2 for India has been reduced by 28.40% from 1990 to 2021 while China has reduced by 86.85%, USA by 74.71% and Russia by 81.17%.

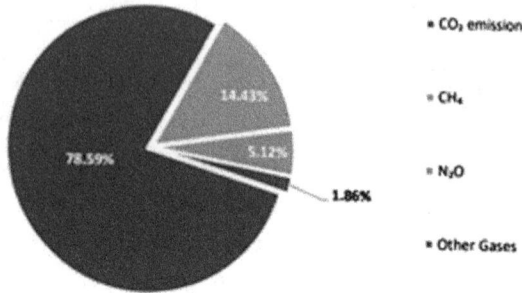

Legend:
- CO_2 emission
- CH_4
- N_2O
- Other Gases

Values shown: 78.59%, 14.43%, 5.12%, 1.86%

Gas-wise emission for the year 2016:

In 2016, the top 15 categories contributing to CO2 emissions are as follows: Electricity Production (40%), Road Transport (9%), Enteric Fermentation (8%), Nonspecific Industries (6%), Iron & Steel (5%), Residential (4%), Cement Production (4%), Agricultural Soils (3%), Refinery (3%), Rice Cultivation (3%), Commercial/Institutional (2%), Cement (2%), Aluminium Production (1%), Lime Production (1%), and Manure Management (1%).

Atmospheric Carbon Dioxide:

According to NOAA's Global Monitoring Lab, global atmospheric carbon dioxide reached a record high of 419.3 ppm in 2023. This marks the 12th consecutive year with an increase of over 2 ppm, a rise of 2.8 ppm from 2022. At Mauna Loa Observatory in Hawaii, the 2023 average was 421.08 ppm.

The rate of CO2 increase has accelerated significantly. In the 1960s, it grew at about 0.8 ppm per year. By the 2010s, this rate tripled to 2.4 ppm per year, making the past 60 years' increase approximately 100 times faster than natural CO2 rises. The more we exceed natural removal processes, the faster CO2 levels rise.

Chapter II

Causes

"Since systematic scientific assessments began in the 1970s, the influence of human activity on the warming of the climate system has evolved from theory to established fact"
Intergovernmental Panel on Climate Change (IPCC)

Climate Change:

Climate change refers to long-term shifts in Earth's local, regional, and global weather patterns. It is driven by an imbalance between the amount of solar energy absorbed by Earth and the amount released back into space. During the day, the Sun's energy enters Earth's atmosphere, and some of this energy is radiated back into space at night.

Since the Industrial Revolution over 200 years ago, human activities have significantly increased the emission of greenhouse gases, such as CO_2, CH_4, CFCs, N_2O, and O_3. These gases trap heat in the atmosphere, leading to a rise in Earth's temperature. Excessive greenhouse gases, primarily CO_2, which is emitted mainly from fossil fuel use, electricity production, transportation, deforestation, and industrial processes, contribute to global warming.

This warming is the primary driver of climate change, disrupting the natural balance of the atmosphere.

Global Warming:

Global warming refers to the long-term increase in Earth's average temperature resulting from human activities, primarily the burning of fossil fuels, which releases greenhouse gases into the atmosphere. Since the pre-industrial period (1850–1900), global temperatures have risen by approximately 1°C, with an average increase of 0.2°C per decade. The majority of this warming since the 1950s is highly likely (over 95% probability) to be caused by human activities, progressing at an unprecedented rate.

Human-induced warming reached approximately 1°C above pre-industrial levels by 2017, with 0.87°C of warming observed between 2006 and 2015. Regional warming has been more pronounced over land than over oceans, with 20% to 40% of the global population experiencing temperatures exceeding 1.5°C during certain seasons.

While past emissions alone are unlikely to push global temperatures to 1.5°C above pre-industrial levels, they have already caused significant changes, including rising sea levels, increased frequency of extreme weather events, and biodiversity loss. Vulnerable populations, particularly in low- and middle-income countries, are experiencing severe impacts such as food insecurity, migration, and poverty.

Scientific consensus indicates that 95% of global warming is driven by human activities, with the remaining 5% attributed to natural causes. Over the past 150 years, human activities have increased atmospheric CO_2 levels from 280 parts per million (ppm) to 421.08 ppm. The Intergovernmental Panel on Climate Change (IPCC) confirms that human-induced greenhouse gas emissions are the primary driver of the observed temperature rise over the last five decades.

Human-made causes

1.   Energy Sector

India's energy policies prioritize ensuring energy security, enhancing access and affordability, diversifying energy sources, improving efficiency, reducing transmission losses, and promoting renewable energy. The power sector relies on conventional sources, such as coal, natural gas, oil, and nuclear power, as well as renewable sources, including wind, solar, hydro, waste, and biofuels.

Energy consumption is the largest source of India's greenhouse gas emissions, accounting for approximately 75%. In 2016, electricity production was the most significant contributor, accounting for approximately 40% of national emissions. Manufacturing and construction industries contributed 18.68% of emissions from the energy sector. By mid-2024, 45.54% of India's installed electricity capacity was from non-fossil fuel sources.

Key contributors to emissions from manufacturing include Iron and steel (43.9%), Cement (13.4%), and smaller shares from industries such as chemicals (0.6%) and Textiles/Leather (1.0%).

In 2023-24, India's per capita electricity consumption increased to 1,395 kWh, a 45.8% increase from 2013-14.

The Indian power sector is driven by state, central, and private players. The total installed capacity in the country is 446,189.721 MW as of June 30, 2024, with the state sector accounting for 24.13%, the central sector for 23.41%, and private producers for 52.46%. Coal contributes 47.28% of electricity, Lignite (1.48%), Gas (5.56%), Diesel (0.13%), Hydro (10.52%), Renewable Energy (33.19%), and Nuclear (1.83%).

2. Industrial Processes and Product Use

Industrial processes, particularly in the mineral, chemical, and metal industries, contribute to greenhouse gas (GHG) emissions. These emissions also arise during the production and use of products, classified under Industrial Processes and Product Use (IPPU), which accounted for 8% of total GHG emissions in 2016.

The primary GHG emitted by the sector is CO_2 (73.4%), followed by carbon tetrafluoride (CF4) (10.67%), hydrofluorocarbons (HFCs) (8.51%), and smaller amounts of methane (CH4), nitrous oxide (N2O), and SF6.

3. Transport Sector

India's transportation system encompasses road transport, railways, aviation, and shipping. Road transport, including buses, trucks, cars, and two-wheelers, is the primary contributor, powered primarily by petrol and diesel, resulting in significant greenhouse gas emissions. Transport accounted for 13% of emissions from the energy sector, with 90% of this coming from road transport, followed by civil aviation (6%), railways (3%), and domestic shipping (1%).

In 2016, emissions from the residential sector (60%) were the most significant contributors outside the transport sector, followed by the commercial sector (32%), and other minor sectors, including agriculture and biomass burning.

India is the world's fifth-largest car market, and as incomes rise and urbanization accelerates, car sales are expected to increase, driving higher global oil demand.

4. Oil and Natural Gas

India's oil consumption accounted for 29% of its total energy use in 2017. Demand is expected to double by 2040, potentially surpassing China as the leading driver of global oil demand growth.

Emissions from oil and natural gas processing occur through fugitive emissions, including equipment leaks, evaporation, venting, flaring, and accidental releases (e.g., pipeline spills or well blowouts). While some emissions are intentional and regulated (e.g., flare systems), others, such as those from production and distribution, involve significant uncertainty in both quantity and composition.

5. Agriculture

Agriculture is crucial to India's economy. Approximately 54.6% of the population is engaged in this activity, contributing 11.05% to the country's GDP. The sector accounts for about 14% of India's greenhouse gas emissions.

Of these, 74% of emissions come from methane produced by livestock (cows, buffalo, goats, and sheep) and rice cultivation. The remaining 26% is due to nitrous oxide (N_2O) in chemical fertilizers. In 2016, agriculture emitted 407,821 Gg of CO_2e, representing 14% of India's total emissions.

Emissions from agriculture include enteric fermentation (54.6%), rice cultivation (17.5%), fertilizer use (19.1%), manure management (6.7%), and field burning (2.2%). N_2O emissions occur through direct and indirect pathways, with direct emissions from fertilizers and indirect emissions from nitrogen volatilization and runoff.

6. Livestock

Livestock plays a key role in India's economy. Two-thirds of the population relies on farming, and India is home to 15% of the world's cattle. In 2014, it produced 19% of the global milk supply, and livestock contributed 4.62% to the country's GDP.

Methane is released during the decomposition of dung, especially in confined animal farming operations such as dairy, swine, and poultry. In 2014, 292.23 million tonnes of dung produced 128.35 Gg of methane emissions. The storage and treatment of dung also generate nitrous oxide (N2O), both directly and indirectly. Dung is used in various ways: 22.5% as dung cakes, 46% in pastures, and 31.5% stored as solid.

7.   Biomass Burning

Biomass burning releases carbon dioxide (CO2), methane (CH4), nitrous oxide (N2O), sulfur dioxide (SO2), and other gases. Crop residue burning, particularly from rice, wheat, cotton, maize, millet, sugarcane, jute, rapeseed, and mustard, is common in India. This practice has a significant impact on air quality, particularly in regions such as Punjab, Haryana, and the NCR, with Delhi being the most affected.

Biomass burning encompasses various activities, including the use of agricultural waste, forest fires, crop residue, and domestic biomass (e.g., for cooking and heating).

8.   Land Use, Land-Use Change, and Forestry

LULUCF activities influence the carbon cycle by altering the balance of carbon stored in vegetation and soils. These activities can either act as carbon sinks, absorbing CO2 from the atmosphere, or as sources of emissions. As human demand for food and infrastructure grows, more land is converted for crop cultivation, roads, buildings, and factories, resulting in increased greenhouse gas emissions.

In 2016, the LULUCF sector served as a net sink, removing 307,820 Gg CO2e, which accounted for approximately

15% of India's carbon emissions. Forests, croplands, and settlements acted as sinks, while grasslands were a net source of CO2.

9. Household Cooking

Nearly half of Indian households still rely on firewood or biomass for cooking, significantly contributing to air pollution. As of 2015, this is responsible for 124,207 premature deaths per one million people. While LPG adoption has increased, with 77 million households receiving connections under the Pradhan Mantri Ujjwala Yojana, many still use biomass in combination with LPG. This leads to 800,000 premature deaths annually due to household air pollution, with women and children most affected. The government aims to use 100% clean cooking fuel by 2030.

10. Rising Population

India's population has grown rapidly, from 23.83 crore in 1901 to over 145 crore in 2024. It is projected to become the world's most populous nation by 2030, with 152 crore people. Population density has risen from 77 people per square kilometre in 1901 to 488 per square kilometre in 2024. High population density puts pressure on resources such as food, water, and energy.

India accounts for 17.78% of the global population, and its annual growth rate is slowing to 0.89%. However, the rising population increases the demand for infrastructure, energy, and agricultural products, contributing to higher greenhouse gas emissions and global warming. More people mean more demand for fossil fuels, which emit carbon dioxide when burned, exacerbating climate change.

11. Urbanization

Urbanization in India is rapidly increasing as more people migrate to cities in search of better opportunities. While only 35% of the population currently lives in urban areas, this is expected to rise to 39.8% by 2030 and 52.2% by 2050. The growth of urban areas leads to a rising demand for housing, transportation, and infrastructure, significantly impacting the environment.

The Indian real estate sector is growing rapidly, contributing to 24% of India's CO2 emissions. Buildings consume approximately 40% of the world's energy, 30% of its raw materials, and generate around 30% of its solid waste. Increasing vehicle use and construction activities worsen air quality and contribute to global warming.

Rapid urbanization also increases vulnerability to climate change, particularly in densely populated cities, with impacts including flooding, rising sea levels, and desertification. While urban growth can drive economic progress, it intensifies fossil fuel use, further accelerating climate change. Sustainable urban infrastructure is crucial to mitigate these effects.

12. Luxury Lifestyles

As India develops and incomes rise, more people are adopting luxury lifestyles, including installing air conditioners (ACs) and refrigerators to cope with extreme heat and preserve food. These appliances, however, emit hydrofluorocarbons (HFCs), potent greenhouse gases with a warming potential significantly greater than that of carbon dioxide, thereby contributing to climate change. They also consume large amounts of electricity, further straining the energy grid and potentially requiring the construction of additional power plants.

Additionally, wealthy individuals and industrialists often use private cars and flights, which consume fossil fuels and release substantial amounts of greenhouse gases into the atmosphere, thereby exacerbating climate change.

13. Waste Sector

The waste sector accounts for approximately 3% of India's total greenhouse gas emissions, primarily due to the decomposition of municipal solid waste and wastewater treatment. Methane (CH_4) is released during the anaerobic breakdown of waste, while nitrous oxide (N_2O) is emitted from domestic wastewater.

India generates the highest amount of waste globally, producing 277 million tonnes of municipal solid waste annually, surpassing China. However, per capita waste is still much lower than that of developed countries. Globally, municipal waste generation is projected to increase by 70%, reaching 3.4 billion tonnes by 2050, unless immediate action is taken, driven by urbanization and population growth.

The National Capital, Delhi, generates the highest waste among all Indian cities, producing 30.6 lakh tonnes annually. It is followed by Mumbai with 24.9 lakh tonnes, Chennai with 18.3 lakh tonnes, Hyderabad with 16.4 lakh tonnes, Bengaluru with 12.8 lakh tonnes, Ahmedabad with 12.1 lakh tonnes, and Pune and Surat, both with 6.2 lakh tonnes. Additionally, there are Kanpur with 5.5 lakh tonnes and Jaipur with 5.4 lakh tonnes.

Bermunda leads the world in waste generation, with 4.54 kg of waste per person per day, followed by the USA at 2.24 kg, Russia at 1.13 kg, and Brazil at 1.04 kg. India generates 0.57 kg of waste per person per day, which is below the global average of 0.71 kg.

## 14.	Forestry Sector

The forestry sector plays a vital role in India's biodiversity. Eighty percent of terrestrial biodiversity is found in forests, and over 300 million people rely on forests for their livelihood. Forests act as crucial carbon sinks, sequestering CO_2 and helping mitigate climate change. In 2016, the Land Use, Land-Use Change, and Forestry (LULUCF) sector sequestered 330.76 million tonnes of CO_2, representing 15% of India's total CO_2 emissions.

According to the Forest Survey of India's report from December 2021, the current forest and tree coverage stands at 24.62%. This includes a forest cover of 21.71% and a tree cover of 2.91%. In comparison, in 2005, the total forest and tree coverage was 23.39%, with forest cover accounting for 20.6% and tree cover for 2.79%. Over the past 16 years, the increase has been just 1.23%.

According to the Global Forest Resources Assessment 2020 (FRA 2020), Forests cover 30.80 percent of the global land area, but are not equally distributed across all countries worldwide. More than half of the world's forests are located in only five countries: Brazil, Canada, China, Russia, and the United States. Two third of the world's forests are found in 10 countries (Russia -20.1%, Brazil-12.2%, Canada–8.5%, USA-7.6%, China–5.4%, Democratic Republic of the Congo-3.3%, Australia–3.1%, Indonesia–2.3%, Peru–1.8%, India-1.8%) and Rest of the World is 33.9%.

According to research data published by a team of 38 scientists in September 2015, the world has 3.04 trillion trees and 422 trees available per person on Earth. Canada has the highest number of trees per person, at 8,953, followed by Russia at 4,461, Australia at 3,266, Brazil at

1,494, the USA at 716, France at 182, China at 102, the UK at 47, and India at 28 trees per person. India has only 28 trees per person, compared to an average of 422 trees per person worldwide.

15. Deforestation

Deforestation in India continues to rise yearly due to industrialization, road expansion, and the construction of residential and commercial buildings. As per the Forest Survey of India's 2021 report, India's forest and tree cover ratio is 24.62%, up from 23.39% in 2005. This marks an increase of only 1.23% despite significant population growth over the past 16 years.

The main causes of deforestation are:

- Agricultural expansion (65%): Crop cultivation and livestock grazing.
- Urbanization (15%): Infrastructure development and housing.
- Infrastructure projects (10%): Roads, dams, and mining.
- Logging and timber extraction (5%).
- Forest fires (5%).

Natural Causes of Climate Change:

Natural factors also contribute to climate change, including the shifting of continents, volcanic eruptions, sea waves, and the Earth's rotation. The Earth's climate takes thousands of years to experience even a one-degree shift in temperature. Volcanic activity releases heat, ash, and gases and can significantly raise atmospheric temperatures. Historical climate changes, such as those during the Ice Age, were influenced by volcanic eruptions, changes in forest ecosystems, solar radiation, comets, meteorites, and other natural events.

1. Volcanism:

Volcanic eruptions have a significant impact on the Earth's climate. During an eruption, vast amounts of sulfur dioxide, water vapor, and ash are released into the atmosphere. When sulfur dioxide interacts with water vapor, it forms sulfuric acid, which can create a global layer of acid in the stratosphere. This sulfuric acid layer absorbs solar radiation, preventing some of the sun's energy from reaching the Earth's surface and cooling the planet. Such eruptions occur several times on average and can result in temporary global cooling effects.

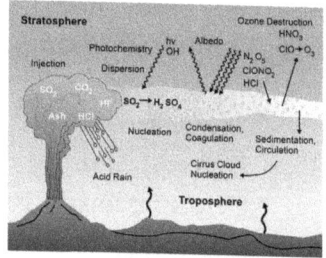

2. Cyclic changes:

Earth's temperature has been fluctuating for many years, and the frequency of natural disasters has been rising. Cyclic changes refer to the natural variations in Earth's systems that occur over different timescales. Some of these changes contribute to climate change, while others may obscure or intensify its impacts.

3. Orbital Variation:

The Earth rotates along its axis at an angle of 23.5 degrees. During the first half of the year, when it's summer, the Northern Hemisphere is

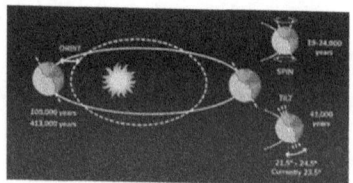

tilted towards the sun. In the second half, the Earth is tilted away from the sun during the colder months. Variations in the Earth's tilt affect the intensity of weather patterns: a more excellent tilt results in more heat and less cold, while a smaller tilt leads to less heat and more cold.

4. Continental Drift:

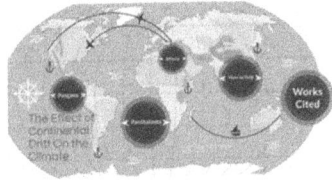

The continents we recognize today were formed hundreds of millions of years ago when landmasses began to break apart. This fragmentation had a significant impact on the climate as it altered the physical characteristics of the land, its position, and the arrangement of oceans. The shifting of continents affected the seas, winds, and wave patterns, influencing the weather. This process of continental drift continues today, as the continents rest on vast tectonic plates that are in constant motion. All these factors contribute to climate change.

5. Sea Waves:

Approximately 71% of the Earth's surface is covered by water, with 96.5% of this water being contained in the oceans. Sea waves absorb heat from the atmosphere and distribute it significantly to the land. Waves transfer heat, momentum, and gases between the ocean and the atmosphere. They also shape coastlines, impacting coastal ecosystems and human settlements. Additionally, waves help regulate the ocean's carbon uptake and storage, contributing to the planet's climate system.

Chapter III

Effects

"The magnitude and rate of climate change and associated risks depend strongly on near-term mitigation and adaptation actions, and projected adverse impacts and related losses and damages escalate with every increment of global warming."
Intergovernmental Panel on Climate Change (IPCC)

1.   Rising Temperatures: The Growing Heat Crisis

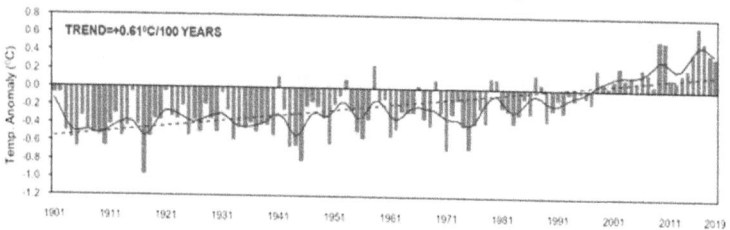

Year after year, surface air temperatures continue to rise, making it increasingly difficult to maintain a normal summer lifestyle. In 2019, India experienced some of its most extreme heatwaves, with temperatures reaching 48°C in Delhi, 50.3°C in Churu, Rajasthan, and 49.2°C in Banda, Uttar Pradesh. The country's average temperature in 2019 was +0.36°C above the 1981-2010 average, making it the seventh warmest year since 1901. Eleven of the 15 warmest years occurred between 2005 and 2019. The past decade was the warmest on record, with a noticeable increase in maximum temperatures. Between 1901 and 2019, India's mean temperature rose by 0.61°C per century. The global average temperature in 2019 was 1.1°C above preindustrial levels. This increase in surface temperatures is primarily driven by rising levels of greenhouse gases, particularly carbon dioxide (CO_2).

2. Intensifying Droughts: A Growing Threat to India's Livelihoods

In 2024, extreme weather events in India became more frequent and intense, with 255 days of such events in the first nine months, compared to 235 in 2023 and 241 in 2022. Many states, including Maharashtra, Karnataka, Andhra Pradesh, Odisha, Gujarat, and Rajasthan, regularly face droughts, with agriculture heavily dependent on the winter rice crop. In 2019, over 42% of India's area experienced drought, affecting around 500 million people. Low rainfall, particularly in southern India (a 47% deficit), exacerbated the situation. Studies show that 60% of districts are unprepared for drought, and only 38% of districts are drought-resilient. Approximately 55 million people globally are affected by droughts annually, threatening their livelihoods and health, and driving migration.

3. Rising Heat Waves: A Growing Crisis for Health, Agriculture, and Productivity

Heat waves, characterized by abnormally high temperatures, occur mainly in India's North-Western regions between March and June, sometimes extending into July. These extreme conditions cause physiological stress and can result in death. Rising temperatures have made heat waves more frequent and intense, significantly impacting human health. In 2018, 19 Indian states were affected by heat waves, up from 9 in 2015. In Odisha, schools close earlier and reopen later due to the heat, which affects education. Agriculture and construction work also face challenges due to the extreme heat. Businesses often restrict outdoor work for safety reasons between 11 am and 4 pm, which reduces productivity. Globally, heatwaves have increased in frequency and intensity, with 125 million more people exposed to them between 2000 and 2016. Between 1998 and 2017, over 166,000 deaths were attributed to heatwaves, including approximately 70,000 during the 2003 European heatwave.

4. Frequent Cyclones: The Growing Threat to India's Coastal Communities

With a coastline of 7,516 km, India is highly vulnerable to natural hazards, especially cyclones, which cause significant losses equivalent to up to 2% of its GDP and 12% of its government revenue. The Indian subcontinent, which is exposed to nearly 10% of the world's tropical cyclones, is frequently affected by cyclones, floods, and other disasters. The most vulnerable states are those along the coast, including Andhra Pradesh, Odisha, Tamil Nadu, West Bengal, and Gujarat. Cyclones occur primarily in May and June, as well as in October and November, with the East Coast being more prone to them due to the higher frequency of cyclones. Between 1980 and 2000, approximately 370 million people were exposed to cyclones each year. The Bay of Bengal generates over 58% of East Coast cyclones. Cyclones cause extensive damage, reversing developmental gains and affecting about 32 crore people, nearly a quarter of India's population.

5. Changes in rainfall Patterns: A Growing Threat to India's Agriculture and Water Scarcity

In recent decades, India has seen irregular rainfall

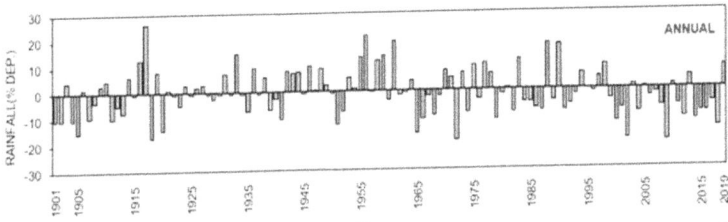

patterns. While the total annual rainfall has remained consistent, the intensity of rainfall has increased, leading to more extreme weather events. Over 75% of the rainfall occurs during the southwest monsoon season (June to September), resulting in water scarcity during the non-monsoon months. The 2019 monsoon saw normal rain, but regional variations occurred, with Central India receiving more than its average. Recent changes have disrupted the seasonal cycle, resulting in intense rainfall, floods, and droughts. For example, Kerala faced severe flooding in 2018 due to abnormally high rainfall, while Bihar, Odisha, Maharashtra, and Karnataka experienced heavy rains and floods in 2019. Global warming has intensified these extreme rainfall events, with heavy rains and floods increasing dramatically worldwide, including a 50% rise in some regions over the last decade.

6. Melting Glaciers and Snowfall: The Threat to South Asia's Water Supply

Glaciers have been melting over the past few decades due to rising temperatures. The Himalayas, home to nearly 22,000 square kilometers of glaciers, are the source of major rivers such as the Ganges, Indus, and Brahmaputra, which provide water to millions of people. Melting glaciers have formed thousands of new lakes, with glaciers retreating at increasing rates—twice as fast in some cases. This accelerated melting contributes to rising sea levels, which have increased by 4-8 inches in the past century. The Hindu

Kush Himalayan region has lost 15% of its glaciers since the 1970s, and projections show that up to 90% could disappear by 2100 if climate change continues unchecked. This will lead to increased flooding and water shortages, resulting in large-scale migration and potential conflict.

7. **Rising Sea Level and Erosion of Beaches: A Threat to India's Coastal Communities**

Sea levels along India's coast have risen by 8.5 cm over the past 50 years, with an average annual increase of 1.7 mm. Coastal areas, such as Diamond Harbour, Kandla, and Haldia, have seen the highest rates of rise. Rising sea levels, caused by warming waters and melting ice, threaten coastal cities such as Mumbai. Extreme sea level events, including storm surges, are expected to become more common by 2100. Global sea levels have risen by 8 inches since 1880 and are projected to rise by another 1-4 feet by 2100. Coastal erosion is also increasing; nearly half of the world's sandy beaches may retreat significantly by century's end, endangering wildlife and coastal settlements. Researchers predict that by 2050, 36,097 km of beaches will be eroded, with even more destruction expected later in the century.

8. **Ocean Acidification: A Growing Threat to Marine Ecosystems and Coastal Livelihood**

Since the Industrial Revolution, humans have released approximately 500 billion metric tons of carbon into the atmosphere, with 30% of it being absorbed by the oceans. This causes ocean acidification, which decreases

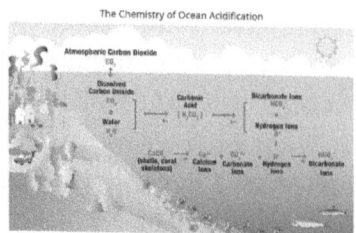
The Chemistry of Ocean Acidification

the pH of seawater and alters the marine carbonate chemistry. From 2009 to 2018, oceans absorbed 22% of annual CO_2 emissions, which also affects ocean chemistry. As a result, the global surface ocean pH has decreased by 0.017–0.027 units per decade, leading to a 26% increase in acidity since the Industrial Revolution. Ocean acidification poses a significant threat to coastal ecosystems, including coral reefs and mangroves, which in turn impact fisheries and the livelihoods of over 400 million people in the Bay of Bengal, where the annual value of the fisheries catch exceeds $1 billion. This also poses risks to the region's tourism and biodiversity.

9. **Shifting of Eco System: Disruption in Plant and Animal Life Cycle**

In India, climate change is causing shifts in ecosystems, disrupting the uniformity of the six seasons. Warm temperatures are arriving earlier in spring and lasting longer in autumn. Human activities, such as the burning of fossil fuels and deforestation, have accelerated climate change, resulting in increased temperatures and altered precipitation patterns, which in turn lead to more extreme heatwaves, droughts, and intense rainfall.

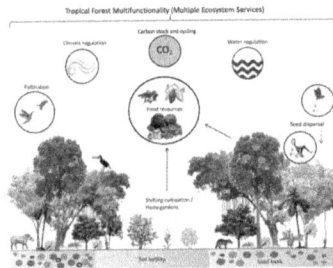

Climate change impacts the environment by raising global sea levels, threatening coastal ecosystems and human livelihoods. It also affects ecosystem services, such as food and water, by altering biological communities and biogeochemical processes. Plant and animal cycles are shifting, with earlier blooms, migrations, and breeding.

Species that migrate between ecosystems are particularly vulnerable to these changes.

10. Health at Risk: The Growing Impact on Public Wellbeing

Climate change affects social and environmental health determinants, including access to clean air, water, food, and shelter, thereby increasing the severity and frequency of existing health issues and creating new ones. Events like the 2013 Uttarakhand floods and the 2004 Tsunami highlight its effects. Globally, environmental risks cause an estimated 12.6 million deaths annually.

Water scarcity, worsened by climate change, increases the risk of waterborne diseases like diarrhoea, which kills 2.2 million people yearly. Climate change also alters vector-borne disease patterns, with India experiencing diseases like malaria and dengue.

Extreme temperatures, such as heatwaves, can lead to heat stress, cardiovascular issues, and respiratory diseases, potentially affecting millions by 2050. Rising temperatures and unpredictable rainfall pose a significant threat to food security, exacerbating malnutrition, particularly in developing countries. Additionally, severe weather events, such as storms and floods, result in death, injuries, displacement, and increased risk of epidemics.

Between 2030 and 2050, climate change is expected to cause 250,000 additional deaths annually due to malnutrition, malaria, diarrhoea, and heat stress.

11. Loss of Economy: The Rising Cost of A Warming India

Climate change is severely affecting the Indian economy. A Stanford study estimates that the country's economy is 31% smaller due to global warming. The agricultural sector, which contributes approximately 16%

to the GDP, struggles with unseasonable rains and frequent droughts, affecting farmers.

Climate change causes permanent economic losses, including decreased crop yields due to hotter weather and drought, costly infrastructure rebuilding after flooding, and higher utility costs for power grids affected by extreme weather. While rebuilding may stimulate short-term economic activity, it doesn't restore productive capacity, such as lost farmland. Investments in infrastructure, however, may improve long-term efficiency.

Climate impacts vary by region and industry, with low-lying, flood-prone areas facing severe risks. A report released ahead of COP28 found that climate change resulted in a 6.3% global GDP loss in 2022, with India experiencing an 8% decline. By 2030, heatwaves could force workers in India to shorten their work hours, risking a loss of up to 4.5% of GDP annually. Climate change could shrink India's economy by 10% by the end of the century, with losses due to reduced labor productivity and rising sea levels. The ADB forecasts a 24.7% GDP loss for India by 2070, with lower-income economies expected to be most affected. According to the Journal of Global Change Biology, labor productivity in India is projected to decline by 40% by 2100.

12. **Farming Under Threat: The Growing Challenges to India's Food Security**

Indian agriculture is highly vulnerable to climate change, particularly due to droughts, floods, heatwaves, and cyclones. Two-thirds of the agricultural land is rain-fed, with irrigation systems relying on monsoon rain. Climate change affects crop growth, soil health, livestock, and pest populations. Increased atmospheric CO_2 benefits crops like wheat and rice, but rising temperatures reduce crop duration, yield, and quality. Extreme weather events and altered rainfall patterns pose a significant threat to agricultural productivity, particularly in rain-fed areas.

Climate change also affects soil quality by decreasing organic matter, increasing soil temperature, and altering erosion patterns. Livestock face challenges from feed quality, water scarcity, and the spread of vector-borne diseases, while rising water temperatures and cyclones impact fisheries. Insects and diseases are likely to spread, affecting crop health.

In India, rice and wheat yields are projected to decrease significantly by 2050 and 2080, with rain-fed crops facing the most substantial reductions. Without adaptation measures, maize yields could also decrease, and overall crop quality may decline due to the effects of climate change.

13. Silent Killers: The Deadly Impact of Environmental Pollution on Health and Economy

Environmental pollution is a significant health hazard, causing millions of premature deaths each year. In 2019, pollution led to over 23 lakh premature deaths in India alone, with air pollution accounting for the majority of these fatalities. Globally, pollution is responsible for approximately nine million deaths, accounting for one in every six deaths worldwide. Air, water, and soil pollution continue to impact public health, causing diseases like heart conditions, lung cancer, and respiratory issues.

Air pollution, primarily from the burning of fossil fuels, has resulted in 12.4 lakh premature deaths in India, reducing global life expectancy by 1.5 years. Water pollution, driven by agricultural runoff and industrial waste, has caused 7 lakh premature deaths in India, and nearly 70% of the country's surface water is deemed unfit for consumption. Due to industrial waste and pesticide overuse, soil pollution hurts crop quality, human health, and ecosystems. These forms of pollution have substantial economic costs, with water pollution alone costing India billions annually. Urgent action is needed to address this invisible crisis, which threatens health, food security, and economic growth.

Chapter IV

Solution

"The scientific evidence is unequivocal: climate change is a threat to human wellbeing and the health of the planet. Any further delay in concerted global action will miss the brief, rapidly closing window to secure a liveable future."

Intergovernmental Panel on Climate Change (IPCC)

The urgency to combat climate change has never been greater. With global temperatures rising and natural resources depleting, taking actionable steps at the individual, community, business, government, and international levels is crucial to mitigate climate change. Here are 365 practical ways and a roadmap for NetZero India by 2050.

Humans are the Drivers of Climate Change and Must Act to Mitigate It

Human activities are the primary cause of climate change as we continue to release large quantities of greenhouse gases into the atmosphere. Mitigating this crisis requires action on multiple fronts.

To prevent the escalating impacts of climate change, greenhouse gas emissions must be drastically reduced. The world is already experiencing the adverse effects of climate change. To avoid even more severe and costly consequences, we must cut emissions by at least 50% by 2030 and achieve net-zero emissions by 2050. The most significant sources of emissions, such as the energy sector, are a good place to begin. However, addressing the full scope of the climate crisis will require comprehensive reductions across all large and small sectors. Key actions include transitioning to renewable energy, electrifying transportation, adopting more sustainable agricultural practices, halting deforestation and forest degradation, increasing reforestation efforts, and minimizing waste production. These measures are critical to reducing emissions and steering the world toward a more sustainable future.

There are three leading solutions to this pressing issue:

- **Limit Carbon Emissions**: Reduce carbon emissions sustainably to the greatest extent possible.
- **Use Renewable Energy**: Generate renewable energy, particularly through rooftop solar panels.
- **Increase Tree Planting**: Plant more trees to offset the carbon in the atmosphere.

However, governments alone cannot solve this problem without the active involvement of every citizen. As individuals, we contribute to the emission of greenhouse gases, so we must take responsibility and work toward the solutions mentioned above, both at a personal level and within our institutions. Simultaneously, the government must implement its climate and environmental policies effectively.

As the old saying goes, "Prevention is better than cure." To safeguard our future, we must collectively strive to reduce carbon emissions at the individual, corporate, government, and international levels in every possible way. *Collective action—by individuals, businesses, and governments at the local, national, and international levels—is essential to combating climate change.*

Conference of the Parties (COP):

"Climate change is a global challenge and it is not a challenge that can be addressed by any one country. We need global cooperation and action."

Dr. Manmohan Singh, *former Prime Minister of India*

The Conference of the Parties (COP) is the highest decision-making body of the United Nations Framework Convention on Climate Change (UNFCCC). This international conference gathers countries from around the world annually to deliberate on and address pressing global climate change challenges. It is a platform for nations to negotiate, coordinate, and implement measures to mitigate climate change and foster global environmental cooperation.

COP objectives:

1. Review and implement the UNFCCC and its Kyoto Protocol
2. Set global climate change policies and goals
3. Establish rules and guidelines for implementing climate change agreements
4. Provide a platform for countries to share experiences and best practices
5. Facilitate international cooperation and climate change negotiations

COP structure:

1. Open to all UNFCCC member countries
2. Presided over by a President, elected by the parties
3. Supported by the UNFCCC Secretariat
4. Meets annually, with each session hosted by a different country

Notable COP outcomes:

1. Paris Agreement (COP 21, 2015): Set a global goal to

limit warming to well below 2°C and pursue efforts to limit it to 1.5°C

2. Kyoto Protocol (COP 3, 1997): Established binding emissions targets for developed countries
3. Copenhagen Accord (COP 15, 2009): Recognized the need to limit warming to 2°C

The COP is crucial in addressing global climate change, promoting international cooperation, and driving climate action.

Paris Agreement:

"The Paris Agreement is a powerful tool to combat climate change, and the United States must lead the world in reducing emissions, fostering clean energy innovation, and creating green jobs."

Joe Biden, former President of the United States

The Paris Agreement, adopted on 12th December 2015 during COP 21, is a landmark global pact under the UNFCCC to combat climate change. It unites all nations to limit global temperature rise to below 2°C, aiming for 1.5°C above pre-industrial levels. The agreement emphasizes support for developing countries, financial resource mobilization, and transparency in tracking progress.

Key Objectives:

- Mitigation: Reduce greenhouse gas (GHG) emissions.
- Adaptation: Enhance resilience to climate impacts.
- Finance: Align funding with low-carbon, climate-resilient pathways.

Features:

- Nationally Determined Contributions (NDCs): Countries submit and update climate action plans.
- Global Stocktake: Conducted every five years to assess progress.

- Financial Goal: USD 100 billion annually through 2025, with new targets from 2025.

Entered into force on 4th November 2016, now ratified by 189 countries.

India's Role and Progress:

India balances economic growth with climate action. Key achievements include:

- 45.54% of electricity capacity from non-fossil fuel sources (2024).
- 33% reduction in emissions intensity of GDP (2005-2019).
- Efforts continue to create an additional carbon sink of 2.5-3 billion tons of CO_2 by 2030.

Updated NDC (2021-2030):

- Reduce emissions intensity by 45% (2005 baseline).
- Achieve 50% non-fossil electricity capacity by 2030.
- Enhance adaptation in vulnerable sectors and mobilize domestic/international funds.
- Commit to net-zero emissions by 2070.
- Creating an additional carbon sink of 2.5 to 3 billion tons of CO2 equivalent by increasing forest and tree cover by 2030.

India demonstrates that climate action and economic development can coexist with a strong focus on sustainable and low-carbon growth.

Individual Level

"Climate change is not just a global issue; it is an issue that affects every individual. We must all act together to fight it."

Narendra Modi, Prime Minister of India

1.   I am NetZero

Net Zero refers to a state where the amount of greenhouse gas (GHG) emissions released into the atmosphere is balanced by the amount of GHG emissions removed or offset. Achieving this balance involves reducing emissions and implementing carbon capture and storage technologies.

As an individual, you have the power to make a meaningful impact by adopting a net-zero lifestyle and encouraging others to follow suit. A Net Zero lifestyle includes:

- Reduce energy consumption: Opt for energy-efficient appliances, switch off lights and electronics when not in use, and replace traditional bulbs with LED or CFL bulbs.
- Use renewable energy: Invest in solar panels or renewable energy credits to power your home.
- Plant trees: Create green spaces around your home, add indoor plants, and develop rooftop gardens to absorb the carbon emissions from your daily activities.
- Eat a plant-based diet: To reduce your environmental impact, choose a diet rich in fruits, vegetables, and whole grains.
- Conserve water: Take shorter showers, fix leaks promptly, and use water-efficient fixtures and appliances.
- Reduce, reuse, and recycle: Minimize waste by reusing items and recycling whenever possible.
- Commute consciously: To reduce your carbon footprint, opt for biking, electric vehicles, public transport, carpooling, or other sustainable modes of transport.
- Buy sustainable products: Choose products made from eco-friendly materials with minimal environmental

impact. Avoid fast fashion and single-use products.

- Educate yourself and raise awareness: Learn about climate change and share your knowledge with others to inspire collective action.
- Support Net Zero policies: Advocate for policies and politicians prioritizing and supporting Net Zero targets.

You can also calculate your carbon footprint at home and in your workplace. You can offset your personal emissions by adopting a Net Zero lifestyle and planting trees. In this way, you can proudly say, "I am NetZero."

Remember, every small action counts. A society full of NetZero individuals can create a significant collective impact, driving change to mitigate climate change. Together, we can make a real difference.

2. Plant a Tree, Shape the Future: Embracing Our Individual Social Responsibility

Human actions have far-reaching effects. With rapid population growth and urbanization, millions of trees are cut annually for infrastructure, food production, and agricultural expansion. India's population has increased by over 100 crore since 1951, yet forest and tree cover have not grown proportionately.

The ideal forest cover is 33% of a country's total geographical area. India's current forest and tree cover stands at 24.62% (2021), far below the global average of 30.8%. If this slow growth rate continues, achieving the government's target of 33% by 2030, as pledged in the UN Climate Change Summit, will be difficult.

The lack of trees worsens greenhouse gas emissions, contributing to climate change and health problems. As individuals, we are responsible for this crisis and its consequences. Just as we support family and friends in times of need, we must now collectively address the shared challenges of climate change and pollution.

Planting trees is a simple yet impactful way to take responsibility. Instead of relying solely on the government or NGOs, we can all invest in tree planting for our own oxygen needs, for future generations, and for the planet. Small individual contributions, such as buying and planting trees from local nurseries, can collectively make a significant difference.

By planting trees, we enhance the environment, beautify our surroundings, and inspire others to act similarly. Together, these efforts can lead to a sustainable future. Every tree counts!

3. Environmental Protection: A Fundamental Duty of Every Indian Citizen

The Indian Constitution preserves environmental protection as a fundamental duty:

- Article 51-A (g) of Indian Constitution says, "It shall be the duty of every citizen of India to protect and improve the natural environment including forests, lakes, rivers and wildlife and to have compassion for living creatures."

- Article 47 provides that the State shall regard the raising of the level of nutrition and the standard of living of its people and the improvement of public health as among its primary duties. The improvement of public health also includes the protection and improvement of the environment without which public health cannot be assured.

- Article 48 deals with the organization of agriculture and animal husbandry. It directs the State to take steps to organize agriculture and animal husbandry on modern and scientific lines. In particular, it should take steps for preserving and improving the breeds and prohibiting the slaughter of cows and calves and other milch and draught cattle.

- Article 48-A of the constitution says that "the state shall endeavor to protect and improve the environment and to safeguard the forests and wildlife of the country".

- Article 21 states, "No person shall be deprived of his life or personal liberty except according to procedure established by law".

Panchayats at the village level are also empowered to take steps for soil conservation, water management, forestry, and ecological preservation.

Environmental protection aligns with India's cultural values, as highlighted in the Atharvaveda: "Man's paradise is on earth; live in harmony with nature." Preserving this paradise is essential for life and well-being.

Let us honor this constitutional duty by actively working to protect and restore our natural resources, just as we united during the COVID-19 pandemic. Together, we can tackle the challenges of climate change and create a sustainable future.

4. Repay the Debt of Nature: Restoring Balance for a Sustainable Future

"Repay the debt of nature" is a powerful and thought-provoking concept that urges us to recognize and address the environmental harm and resource depletion caused by human activities. It is a call to restore balance between our needs and the Earth's natural limits.

We take oxygen from the environment daily,

primarily from trees. Our ancestors and forefathers planted the forests and trees that provide this oxygen. If we continue to use oxygen without planting new trees, oxygen supplies will soon be depleted, and we, along with all other living creatures, will cease to exist.

Just as we borrow money from banks, friends, or relatives and repay it over time, nothing in life is truly free. For shelter, we pay for purchasing homes, annual ground rent, and property taxes. For food, we buy groceries, vegetables, and cooking supplies. For clothing, we purchase items from stores. We pay for luxuries like air conditioners, refrigerators, televisions, cars, and mobile phones. Even for entertainment, whether it's movies or sports, we spend. For electricity, we pay monthly utility bills. For health and fitness, we invest in gyms, yoga centres, and healthcare services when sick. We also pay for local transport, rail, air, or personal vehicles.

Water, once considered a free resource, is no longer free. We pay for municipal water taxes; for drinking water, we purchase bottled water or install water purifiers in our homes. The price of water, which ranges from Rs. 15 to Rs. 25 per litre, varies depending on the brand.

Now, consider oxygen — is it free? No. While we breathe it freely, we only have access to oxygen because of the trees planted by our ancestors, and the government spends thousands of crores annually to plant and protect forests. In hospitals, oxygen comes at a significant cost, especially when supplied in cylinders. Thus, we are constantly borrowing oxygen from nature, oxygen provided by trees planted long before us, and is currently being sustained by government investments for forest conservation.

If we can repay debts like loans, credit card bills, or personal borrowings, why not the debt to nature?

It is our responsibility to repay this debt by planting and protecting trees. We need to give back the oxygen we've consumed since birth by planting new trees and ensuring they thrive throughout our lives. Just as we nurture our children until they are self-sufficient, we must care for newly planted trees for at least three years to ensure their survival. This is as important as the care we give to our own children.

While we save for our children, trees provide for us throughout their lives: oxygen, fruits, shade, wood, shelter for wildlife, and medicinal properties from their leaves and bark. Trees ask for nothing in return but offer everything they have, from birth to death.

We must think about better caring for the trees around us, so they can continue to serve us. The more trees there are in your surroundings, the more oxygen you have to breathe. Clean air means better health for your family, friends, community, and society.

In ancient times, our ancestors revered and worshipped trees. Today, we continue honoring sacred trees like the Peepal, Bael, and Tulsi plants. So, plant trees, nurture those around you, and protect the environment to repay nature's debt.

Make conscious choices in your daily life to reduce your ecological footprint and contribute to a sustainable future. Remember, repaying the debt to nature is a collective responsibility that requires long-term dedication to environmental stewardship.

Let's all work together to repay our debt to nature and create a more sustainable and prosperous future for future generations!

5.   Forest Officials: Key Driving Force in Forest Conservation

Forest officials are pivotal in protecting and managing forests. Their key responsibilities include:

- Enforcing forest laws and regulations.
- Monitoring forest health and addressing threats like pests or diseases.
- Implementing sustainable forest management practices.
- Collaborating with communities and stakeholders.
- Promoting public awareness about forest conservation.

Forest officials work to conserve natural resources and ensure ecological balance under the guidance of the Ministry of Environment, Forest, and Climate Change (MoEFCC). At the state level, Forest Departments, led by Indian Forest Service (IFS) officers, implement policies to enhance forest cover and maintain ecological stability.

Despite efforts, India's forest and tree cover is only 24.62% (as of 2021) against the target of 33% set in 1950. Challenges include inadequate coordination with local communities, poor monitoring of sapling survival, illegal tree cutting, and weak enforcement of environmental laws.

Forest officials must strengthen community engagement, enforce laws effectively, and promote accountability to achieve the 33% target. Public support and encouragement for their efforts are vital to mitigating climate change and ensuring a healthier environment for future generations.

6. Block and District Administrative Officers: Catalysts for Environmental Protection

Approximately 65% of India's population resides in rural areas. Block and District administrative officers like Block Development Officers (BDOs), Tahasildars, District Collectors, Sub-Collectors, and Deputy Collectors are highly respected figures. These officials hold significant authority and influence, making them pivotal in driving local development, including environmental protection.

Their Role in Environmental Conservation:

- Influential Leadership: Through their roles in rural development, they can promote environmental initiatives as part of holistic growth.
- Constitutional Duty: Article 51-A (g) emphasizes that environmental protection is a fundamental duty for all, including administrative officers.

Practical Initiatives:

Block-Level Actions:

- During Panchayat meetings, BDOs can advocate for environmental protection, encourage tree plantations, and foster climate awareness.
- Facilitate government schemes, such as promoting LED bulbs, solar-powered streetlights, solar lamps, and household toilets under the Swachh Bharat Mission.
- Address waste management and support village cleanliness drives, including pond restoration.
- Promote fruit tree plantations by households on vacant land.

District-Level Leadership:

- Collectors and Sub-Collectors can highlight climate change impacts during district and block-level interactions.
- Conduct surprise village visits to assess ground realities and improve environmental protection measures.
- Align available government schemes with local needs and encourage community participation.

Why Their Role Matters:

Block and district Officials Bridge the gap between

government policies and grassroots implementation. By leveraging their influence and authority, they can inspire communities to adopt sustainable practices and address climate change.

Encouraging and supporting these officers in their initiatives can significantly enhance local climate action and environmental conservation efforts!

7. Police Officials: Enforcers of Environmental Laws

Police officials play a vital role in maintaining law and order, protecting property, and ensuring public safety. However, their involvement in enforcing environmental laws remains limited.

Current Scenario:

- Police stations primarily handle criminal cases, with few environmental violation cases being reported or filed.
- Many citizens prioritize immediate concerns over environmental issues, leading to a lack of reporting on environmental violations.

Potential Role of Police:

- Encouraging Reporting: Police can motivate the public to file complaints against environmental law violators, raising awareness about the importance of protecting the environment.
- Awareness Campaigns: Police officials can educate communities about environmental protection and its impact on public health by participating in and conducting awareness programs.
- Collaboration with Authorities: Work closely with forest departments, municipal corporations, and pollution control boards to address violations effectively.

Impact:

Increased filing of environmental violation cases will deter offenders and foster a culture of accountability and environmental responsibility. Police officials can be key contributors to combating climate change and promoting sustainability.

Let's support and encourage police officials to take active steps in enforcing environmental laws and protecting our planet!

8. Doctors: Not Just Lifesavers, But Guardians of Health and the Environment

Doctors are highly respected in Indian society. They are often viewed as God-like figures for their ability to save lives. They serve as role models for many students, and countless parents aspire for their children to join the noble medical profession and contribute to society.

Beyond their medical expertise, doctors can also play a pivotal role in environmental conservation. By addressing the health impacts of climate change and advocating for sustainable healthcare practices, they can protect both human health and the environment.

Let's honor and appreciate doctors who lead the way in championing environmental causes and inspiring others to do the same! They are often viewed as God-like figures for their ability to save lives. They serve as role models for many students, and countless parents aspire.

9. Political Representatives: Game-Changers for Environmental Protection

Political leaders from ruling or opposition parties hold significant influence, especially in India, where people actively listen to and follow them. Their influence is even more significant in rural areas. Regular party meetings, social media outreach, and election rallies provide them vast platforms to address pressing issues.

Unfortunately, environmental protection and climate change are rarely prioritized in political arenas, where people actively listen to and follow them. Their influence is even more significant in rural areas, where topics like unemployment, infrastructure, and welfare schemes overshadow them. While some parties have begun including renewable energy and sanitation in their manifestos, the urgency of climate change remains underemphasized.

Steps Political Leaders Can Take:

- In Party Meetings: Incorporate discussions on environmental protection, promoting practices like tree plantation, solar energy adoption, LED usage, and Swachh Bharat initiatives.
- Tree Plantation Drives: Encourage party workers to plant and care for trees during gatherings, spreading community awareness.
- Municipal & Panchayat Levels: Advocate for clean energy solutions, better waste management, and village greenery in local meetings.
- Election Rallies & Manifestos: Highlight environmental issues and climate action as key agendas, inspiring public participation and awareness.

A single appeal from influential leaders can create a massive ground-level impact. For instance, PM Narendra Modi's "Ek Ped Maa Ke Naam" campaign inspired millions to plant trees in honor of their mothers, showcasing the power of leadership in driving change.

Political representatives have the potential to transform environmental awareness and action. Let's support and encourage them to prioritize sustainability, ensuring a healthier planet for generations to come!

10. Adoption of Villages: A Path to Sustainable Development

Village adoption is a proven approach for creating lasting change, with governments, banks, businesses, and celebrities leading efforts. It focuses on sustainable and inclusive growth, empowering communities to achieve development aligned with Sustainable Development Goals (SDGs).

Key Example:

The Sansad Adarsh Gram Yojana (SAGY), launched by PM Narendra Modi in 2014, encourages Members of Parliament to adopt villages and transform them into model villages with innovative schools, basic healthcare, housing, and sustainable infrastructure.

Steps for Scaling Adoption:

- State-Level Initiatives: MLAs, both ruling and opposition, can adopt villages in their constituencies, showcasing progress during campaigns.
- Comprehensive Development: Adopted villages can focus on greenery, clean energy, waste management, 100% toilets, clean water access, solar energy, and income generation.
- Corporate and Celebrity Involvement: Individuals and organizations can adopt villages for modern, eco-friendly transformations beyond CSR.

Ongoing Efforts:

Banks like NABARD are already driving financial and developmental initiatives in villages. Expanding such efforts will amplify sustainable development across the country.

Village adoption empowers communities, fosters sustainability, and mitigates climate change. Scaling this model nationwide can significantly enhance rural living and environmental protection!

11. Celebrities: India's Role Models for Change

In India, celebrities such as film stars, athletes, and artists command massive followings, with millions of fans adopting their styles and listening to their messages. These influencers have a unique ability to rally public attention and support for critical issues.

Contributions to Social Causes:

Celebrities often donate to and campaign for causes like disaster relief and public health. Similarly, they can champion environmental protection, a pressing global concern, by:

- Creating Awareness: Advocating tree plantations, renewable energy, and sustainable living.
- Setting Examples: Riding bicycles, adopting villages for development, and promoting emission-free initiatives.
- Fundraising: Supporting environmental projects through donations and campaigns.

Notable Environmental Advocates:

- Priyanka Chopra (UNICEF Goodwill Ambassador)
- Aamir Khan (Water Conservation)
- Dia Mirza (UN Environment Goodwill Ambassador)
- Akshay Kumar (Sanitation)
- Nagarjuna (Forest Conservation)

Impact Potential:

Celebrities can inspire millions to take climate action, transforming awareness into tangible environmental benefits. Their efforts can lead India toward a sustainable future.

Let's encourage more celebrities to join the fight for our planet and motivate their fans to do the same!

12. Teachers: The Ultimate Gurus

Teachers hold a unique and enduring place of respect in Indian society. Unlike many professions, their influence and honor extend far beyond retirement, as they are seen as the creators of future leaders like IAS, IPS, and IFS officers.

Shaping Future Generations:

Teachers and lecturers are pillars of education, guiding students who consider them role models. Beyond academics, they instill values and teach extracurricular skills, including environmental awareness, sports, arts, and yoga.

Role in Environmental Protection:

In the face of climate change, teachers can inspire students to:

- Engage in tree plantation and care.
- Adopt sustainable, low-emission lifestyles.
- Take environmental protection initiatives in schools, homes, and communities.

Teachers nurture future climate leaders by empowering students with the knowledge and values of sustainability, contributing to a greener, more sustainable India.

Let's honor and support our teachers as the ultim0061te Gurus shaping a Net Zero India by 2050!

13. Panchayat Sarpanch: Key Leader for Ground-Level Change

The Gram Panchayat, a vital local government institution, is responsible for the development and welfare of villages. Under the Seventy-third Constitutional Amendment Act, 1992, Gram Panchayats are entrusted with 29 functions, including social forestry, water management, health, sanitation, agriculture, and renewable energy, making the Panchayat Sarpanch a crucial figure in ensuring environmental protection.

Role and Responsibilities:

1. Utilizing Panchayat Development Funds:
   - Plant fruit and medicinal trees in vacant lands for sustainable income and beautify roads with flower trees.
   - Promote greenery around Panchayat offices as an example for households.
2. Promoting Renewable Energy:
   - Encourage households to install solar panels, use solar lamps, and replace traditional bulbs with energy-efficient LED bulbs.
3. Environmental Awareness Campaigns:
   - Organize camps with experts like forest officials, doctors, and administrators.
   - Reward individuals contributing to environmental efforts with prizes and certificates, fostering healthy competition.

A proactive Panchayat Sarpanch can lead the community toward sustainable practices, reduce carbon emissions, and enhance resilience against climate change.

By leveraging their position and resources, Sarpanchs can transform villages into sustainable and environmentally conscious communities, setting an example for the rest of the country.

14. Bring Nature to Your Home for Peace and Wellness

Integrating nature into your home can significantly enhance your well-being, reduce stress, and create a harmonious atmosphere for your family.

Benefits of Bringing Nature to Home:
- Improves Air Quality: Plants like Aloe Vera and Peace

Lily absorb pollutants and release oxygen, making indoor air cleaner.

- Regulates Temperature and Humidity: Greenery helps balance indoor temperatures and maintain humidity.
- Reduces Stress and Anxiety: Nature creates a tranquil environment, promoting relaxation and mental peace.
- Boosts Productivity: Adding plants to workplaces enhances focus, creativity, and productivity.
- Enhances Aesthetic Appeal: Indoor plants bring beauty, freshness, and positive energy to your space.
- Encourages Positive Relationships: Nature fosters a calm environment, easing tensions and strengthening bonds.

Simple Steps to Bring Nature Home

- Choose Indoor Plants: Add plants like Money Plant, Snake Plant, Lucky Bamboo, or Tulsi to your living spaces for their air-purifying and stress-relieving properties.
- Balcony and Surroundings: Plant Tulsi and Neem in your balcony or backyard. Surround your home with flowering and fruit trees like Mango, Neem, and Amla.
- Incorporate Greenery in Workspaces: Place small indoor plants on desks to enhance focus and reduce anxiety.

By bringing nature to your Home, you create a sustainable and peaceful environment that benefits your health, supports biodiversity, and contributes to climate resilience. A greener home is a happier home!

15. Greening Rural India: Economic Growth Through Fruit Tree Plantations

In India, where 65% of the population resides in rural areas, vast vacant land remains underutilized. Despite numerous welfare schemes by the government, such as free ration, housing, and cash transfers, these initiatives have unintentionally reduced villagers' motivation to work. However, horticulture and agriculture schemes present a significant opportunity for rural development. Villagers can use their vacant or community land to plant fruit trees like mango, lemon, guava, cashew, and jackfruit, generating additional income and enhancing land productivity.

Tribal areas, particularly those with wastelands, can benefit from targeted initiatives like NABARD's Wadi project. This program supports tribal families in cultivating one-acre fruit plantations, providing additional soil and water conservation resources, sustainable agriculture, and training. Families earn ₹50,000–₹60,000 annually from these plantations and seasonal crop income. Such initiatives can be further promoted through Corporate Social Responsibility (CSR) funds and awareness campaigns led by the Horticulture Department.

Besides boosting household incomes, fruit tree plantations contribute to environmental sustainability by absorbing greenhouse gases. A nationwide effort to convert vacant lands into productive orchards can drive economic growth, create jobs, and support climate change mitigation.

16. Green Roads, Healthy Villages: Empowering Rural Communities through Roadside Tree Plantations

Approximately 65% of Indian roads are in villages, where most of the rural population resides. Although the government, through forest departments and NGOs, undertakes roadside plantation projects, much of the potential remains untapped. Due to limited funding and competing priorities, relying solely on government initiatives is insufficient.

Communities must act proactively as climate change causes droughts, heatwaves, floods, and pollution-related health issues. Trees, the "unsung heroes," absorb greenhouse gases and combat climate change. Villagers can take the initiative to plant fruit trees like mango, guava, jackfruit, and jamun, along with medicinal plants like bael, amla, and lemon, along village roads. These trees beautify the environment and provide fruits, economic benefits, and health advantages to the community.

Planting and caring for roadside trees can significantly contribute to a greener, healthier, and more sustainable rural environment, empowering communities to mitigate climate change actively.

17. Urban Greenery: Planting Trees to Revive City Roads and Combat Pollution

With 35% of India's population living in urban areas, cities continue to grow rapidly due to better income opportunities and infrastructure. However, this urban expansion has increased greenhouse gas emissions, worsening air quality and the environment.

While suffering the effects of climate change and pollution, many city residents often view tree plantation as solely the government's responsibility, reasoning that taxes already fund for environmental initiatives. However, combating climate crises requires individual effort. Waiting for government action alone is insufficient; city dwellers must take the initiative to plant trees along urban roads to protect their environment and themselves.

Tree-lined roads in New Delhi stand out as a model of excellence in combining architectural aesthetics with environmental benefits. Planting trees enhances urban beauty, absorbs greenhouse gases, improves air quality, and contributes to a healthier, greener city.

By actively participating in roadside tree plantations, urban communities can be crucial in mitigating climate change and creating a sustainable urban environment for future generations.

18. Planting Trees in Marketplaces: Creating Greener, Healthier Urban Spaces

Urban marketplaces in India are often congested, overcrowded, and plagued by poor air quality, making them uncomfortable for shoppers and traders, especially during the summer. Planting trees in these areas can help cool the environment by absorbing carbon dioxide and improving air quality.

Market associations, comprising shopkeepers and business owners, can lead tree plantation efforts by utilizing available spaces in and around their complexes. These associations, along with individuals, can ensure the growth and maintenance of the trees.

Planting trees in urban marketplaces improves air quality, carbon sequestration, and provides shade to reduce heat, noise reduction, enhances mental well-being, increases biodiversity, provides aesthetic appeal, and increases resilience to extreme weather. Tree planting also promotes sustainable urban planning and fosters community engagement.

Planting trees in marketplaces can create greener, healthier, and more vibrant community spaces while contributing to climate change mitigation.

19. Green Pride: Celebrating National Holidays with Tree Planting

National holidays like Independence Day, Republic Day, and Gandhi Jayanti offer a perfect opportunity to honor the environment. By planting trees these days, we can symbolize our commitment to protecting nature while celebrating our nation's heritage. Tree planting initiatives can unite communities, promote sustainability, and create lasting green legacies for future generations.

I celebrate National Holidays in our colony yearly by planting trees and hosting a post-flag. People are showing more interest in tree planting, taking pictures to share on their social media, and getting good appreciation for their posts. Also, they feel happy when they see the growth of their plants. People are taking pictures with single trees, families, and groups with trees. That means tree planting gives them immense pleasure, so they take multiple pics in different poses along with multiple individuals and share them on their social media.

Let's celebrate national pride by planting trees for a healthier, greener India.

20. Celebrate with Trees: Creating Lasting Memories for the Planet

As income levels rise, celebrations in India have become impressive, with people marking birthdays, anniversaries, festivals, and community events through parties, rituals, and extravagant spending. However, such celebrations often harm the environment, leaving many unfulfilled in their pursuit of happiness. Instead of giving up celebrations, we can adopt sustainable practices that balance joy with environmental responsibility.

Every celebration —birthdays, weddings, and

anniversaries—presents an opportunity to plant trees as a lasting memory. Cutting costs from lavish celebrations to purchase and plant trees—in homes, societies, or villages—can create meaningful, eco-friendly traditions and leave a legacy that grows with time while promoting environmental awareness.

My wife was pleased and happy when she saw the growth of trees planted on our 10th wedding anniversary in May 2020. Instead of spending money at parties or giving gifts to the wife, 100 Kadam and Badam trees were planted at Devendra Nagar, Raipur, Chhattisgarh.

One of my friend and Ex-office colleagues, Sri Narendra Singh Thakur, felt happy when he saw the growth of trees planted on their 8th marriage anniversary. Mr Thakur and his wife planted 8 trees at Gurughasidas Garden, Sector-3, Raipur, Chhattisgarh.

Every tree planted during a celebration becomes a living symbol of joy and commitment to a sustainable future. Integrating tree planting into our festivities can strengthen our connection with nature and leave a lasting legacy for generations to come.

21. Honoring Loved Ones: A Living Tribute Through Tree Planting

Planting trees to remember a loved one is a beautiful and meaningful tribute. Trees are symbols of life, growth, and resilience, and they can provide a lasting, living reminder of someone special. Many people choose to plant trees in parks, gardens, or even through organizations that plant trees in forests to honor loved ones.

Have you ever considered planting a tree for someone, or is this something you're thinking about doing?

In July 2016, I gifted 500 hybrid Mango trees to 500 families in memory of my maternal uncle, Late Shri Dhaneswar Singh, in his village and neighboring villages, where people were happy and appreciated my initiative. But during that time, my father was not ready to gift trees on the budget cost of giving food to villages and Brahman Bhojan due to the mindset of people. Later, I managed this effort by spending extra budget.

In October 2024, I gifted 300 hybrid Mango trees to all the guests who came on my mother's 1st Death Anniversary. People have shown more interest in taking a tree instead of showing less interest in foods served for lunch arrangements.

22. Gift a Tree: A Meaningful Present for Loved Ones and the Planet

A meaningful gift creates a lasting memory and symbolizes your love and gratitude. Unlike material gifts often forgotten in storage, gifting trees offers a lifetime of value for your loved ones and the environment.

We often spend considerable time and money selecting gifts for birthdays, anniversaries, Friendship Day, Valentine's Day, and festivals like Diwali, Christmas, and New Year. However, these gifts are often underutilized or forgotten after a few years. Instead, consider gifting trees—a unique and impactful gesture.

Plant trees in the name of your loved ones, capture their growth through photos, and share these moments. This gift will serve as a lifelong memory and a symbol of your care. Trees bring happiness and peace and benefit society by contributing to a greener environment.

Encourage family, friends, and relatives to join this meaningful trend. If every Indian gifted trees, the nation could achieve its goal of increasing forest cover to 33% by 2030, fulfilling its carbon sink commitment of 2.5–3 billion tonnes of CO_2 equivalent.

For those unable to plant trees themselves, NGOs and plantation companies offer services to plant trees on your behalf and provide tree certificates as proof of your contribution.

By gifting trees, you inspire loved ones to cherish nature, combat climate change, and create a sustainable future—all while expressing your care thoughtfully and enduringly.

23. Green Your Parking Lot: Planting Trees for a Sustainable Urban Future

In many cities, trees are often removed to make way for parking lots, whether for residential or commercial purposes. However, planting trees in and around parking areas offers numerous benefits for both the environment and the community:

- Carbon Sequestration: Trees absorb CO_2, reducing greenhouse gas emissions.
- Heat Reduction: Shaded parking areas remain cooler, mitigating the urban heat island effect.
- Air Quality Improvement: Trees filter pollutants, improving air quality and public health.
- Storm water Management: Tree roots absorb rainwater, reducing surface runoff and preventing flooding.
- Biodiversity Support: Trees provide birds, insects, and other urban wildlife habitats.
- Soil Stabilization: Tree roots help prevent soil erosion in paved areas.
- Mental Well-being: Green spaces around parking lots reduce stress and enhance well-being.

- Enhanced Aesthetics and Property Value: Landscaping with trees improves properties' visual appeal and market value.
- Climate Resilience: Trees contribute to global efforts to combat climate change.

By integrating tree planting into parking lot designs, we can create greener, cooler, and more sustainable urban spaces while supporting global environmental goals. This simple step transforms parking lots into eco-friendly spaces that benefit everyone.

24. Greening Highways: Transforming Roads into Eco-Friendly Pathways

India boasts an extensive road network of 6,331,791 kilometers as of December 2022, including 146,126 kilometers of national highways as of July 2024. These highways connect rural villages and urban cities, making them crucial for transportation. Enhancing local highways with greenery can significantly benefit the environment and communities.

Although the government has initiated tree plantation projects along highways, public participation is essential to amplify these efforts. Individuals or community groups can take the initiative to plant trees along the highways near their villages or cities, contributing to a healthier environment and a greener future.

Benefits of Highway Greenery:

- Carbon Sequestration: Trees absorb CO_2, reducing greenhouse gases.
- Oxygen Production: Greenery enhances air quality by releasing oxygen.
- Heat Reduction: Vegetation helps cool the area, mitigating the urban heat island effect.

- Biodiversity Support: Green spaces provide habitats for wildlife.
- Pollution Control: Trees filter pollutants, improving public health.
- Aesthetic Enhancement: Green highways create visually appealing landscapes, boosting community pride.
- Noise Reduction: Vegetation acts as a natural sound barrier.
- Storm water Management: Greenery aids in controlling water runoff and supports sustainable drainage.
- Eco-Awareness: Green highways inspire environmentally friendly practices.
- Global Impact: Such initiatives support global climate change mitigation efforts.

Greening your local highways contributes to a sustainable and eco-friendly environment while creating a more beautiful and inviting space for all. This collective effort can make a significant difference in combating climate change and improving community well-being.

25. Greening Your Balcony: A Simple Step for a Healthier Planet

Growing plants on your balcony offers multiple benefits, including helping to mitigate climate change. Plants in balconies contribute to air purification by absorbing CO_2, pollutants, and toxins while releasing oxygen to improve air quality. They also help regulate temperature by providing shade, reducing urban heat islands, and storing carbon in their biomass and soil, aiding carbon sequestration.

Additionally, balcony plants create habitats that support local biodiversity. You can choose easy-to-grow options such as herbs like basil, mint, and rosemary; leafy greens like lettuce, spinach, and kale; flowering plants like petunias, marigolds, and sunflowers; or fruiting plants such as tomatoes, strawberries, and citrus.

Beyond their environmental impact, balcony plants enhance mental well-being, provide fresh produce and herbs, and improve the aesthetic appeal of your home.

Start your balcony garden today and create a greener, healthier environment while enjoying the beauty and benefits of nature!

26. Transform Your Terrace into a Green Oasis

Terrace gardening has become popular for urban residents and those with limited outdoor space. It offers a convenient way to grow plants and enjoy the rewards of gardening. Terrace gardening also adds decorative and functional value to outdoor spaces and provides a fulfilling experience for those who enjoy nurturing plants.

Ideal Terrace Plants:

- Fruit Trees: Citrus, Mango, Guava
- Vegetable Gardens: Tomatoes, Cucumbers, Carrots
- Herb Gardens: Basil, Mint, Rosemary
- Flowering Plants: Marigolds, Sunflowers, Petunias
- Succulents: Aloe, Echeveria, Crassula

Benefits of Terrace Gardening:

- Access to fresh produce and herbs
- Improved mental health and well-being
- Enhanced aesthetic appeal of your home
- Increased property value
- Opportunities for community engagement

Cultivating plants on your terrace contributes to climate change mitigation, improves air quality, supports local biodiversity, and enhances urban aesthetics. Start terrace gardening today to create a sustainable and rejuvenating green space while making a lasting impact!

27. Planting Trees: A Path to Abundant Blessings

God has bestowed humans with intelligence, the ability to think, and the capacity to distinguish right from wrong. The blessings we receive reflect our actions—good deeds bring blessings, while wrongdoings result in negative consequences. Visiting places of worship and offering prayers for blessings is common, but blessings alone cannot replace the fruits of hard work. Simply visiting a temple without effort will not yield results; instead, sincere work combined with prayer fosters peace of mind and fulfillment.

Helping the needy, supporting orphanages, aiding old age homes, and donating food or clothes are noble actions that benefit individuals and bring blessings. However, planting trees multiplies these blessings manifold, as trees provide benefits not just to individuals but to entire communities. The oxygen, shade, fruits, and habitat for birds and animals that trees provide help thousands, even millions, over generations.

A priest once said, "The blessings earned from planting a single tree can overshadow the wrongs of ten misdeeds." Trees contribute to society in countless ways—offering clean air, medicinal benefits, shelter, and sustenance. While helping individuals creates visible, immediate results, trees' intangible, long-lasting benefits extend across time and space, benefiting humanity and nature alike.

The universal law of attraction teaches that the more people benefit from your actions, the more blessings you attract. Planting trees nurtures this principle by benefiting both current and future generations. You can enhance these blessings by:

- Planting and caring for trees with love and dedication.
- Spending time with trees, appreciating their shade and beauty.
- Practicing mindfulness, meditation, or yoga under trees.
- Showing gratitude and respect for trees through rituals or ceremonies.
- Supporting reforestation efforts and using eco-friendly products.
- Inspiring others to plant trees and promoting sustainability.
- Creating art or literature inspired by trees and their wisdom.

Deepening your connection with trees fosters a harmonious relationship with nature and amplifies the blessings you receive. By planting trees, you contribute to a healthier, more sustainable world, ensuring long-term benefits for countless beings.

Plant more trees—bless others, bless the planet, and receive abundant blessings in return!

28. Solar Rooftops: Paving the Way for a Greener Future

Solar energy is crucial for reducing greenhouse gas emissions generated by traditional coal-based power plants. Rooftop solar systems have gained popularity as a renewable and sustainable source of electricity, offering a consistent power supply with minimal recurring costs. With proper planning and installation, these systems provide an eco-friendly energy solution that is both practical and efficient.

In India, rural and urban areas offer millions of rooftops that receive abundant sunlight throughout the day, making them ideal for solar energy generation. Solar panels installed on these rooftops convert sunlight into direct current (DC) power, which is then transformed into alternating current (AC) through an inverter to power household appliances and devices.

A personal experience highlights the importance of rooftop solar systems. During Cyclone FANI in May 2019, Puri district in Odisha experienced extended power outages—15 days in town areas and up to two months in villages. My family, including a young child, struggled in the hot summer nights without electricity. A rooftop solar system purchased by my brother-in-law became a lifeline, powering fans and lights for 6-8 hours daily during the crisis. However, after electricity was restored, the system was stored away—a reflection of the limited awareness and comfort-driven mind-set that often hinders widespread adoption of solar energy.

The Central and State Governments offer various subsidy schemes for residential and commercial solar installations to encourage solar energy adoption. Despite these incentives, many people hesitate to invest in rooftop solar panels due to the initial high costs, long-term savings horizon (10-25 years), and a lack of understanding about the urgency of climate change and the role of renewable energy.

Installing rooftop solar panels empowers individuals with sustainable energy solutions and contributes to a cleaner, greener planet. By reducing reliance on conventional energy sources, we can combat climate change while ensuring energy security for future generations.

Let's embrace rooftop solar systems to create a brighter, cleaner energy future for our homes, communities, and the planet!

29. Empowering Rural India: Solar Lamp Distribution for a Sustainable Future

Solar lamps provide an effective solution in rural areas of India, where access to electricity is limited and power cuts are frequent. Many rural communities still use kerosene lamps for lighting during power outages, posing significant health and environmental risks. Solar lights are crucial for households without electricity connections or frequent power disruptions.

Solar-powered lamps are not only cost-effective but also healthier and environmentally friendly. They are up to 10 to 20 times brighter than kerosene lamps, eliminating the risks of accidental fires and reducing indoor pollution caused by harmful smoke. Kerosene usage is limited and insufficient for extended lighting periods, whereas solar lamps provide consistent illumination without causing health hazards. The environmental impact of kerosene use, which contributes to respiratory diseases and annual fatalities, highlights the importance of transitioning to solar-powered solutions.

During Cyclone FANI in May 2019, Odisha faced widespread power outages lasting 15 days to 2 months. Solar lamps were critical in lighting homes during this crisis, and supplies were distributed from neighboring states to meet the demand. This highlights the necessity of sustainable solar solutions in rural regions.

The distribution of solar lamps in rural India helps mitigate climate change by reducing reliance on fossil fuels, decreasing greenhouse gas emissions, conserving energy, and supporting sustainable livelihoods. Expanding access to solar-powered lighting can contribute to global climate change mitigation efforts and promote a cleaner, greener future for rural communities.

Distribute solar lamps in rural India to foster sustainable energy access and protect the environment!

30. Switching to LED: A Bright Step Towards Energy Efficiency and Sustainability

Energy efficiency plays a crucial role in addressing the nation's electricity shortages and concerns related to climate change. Lighting alone accounts for about 20% of India's total electricity consumption.

While rural electrification has made significant strides, power cuts remain a persistent issue. To meet lighting needs, many households rely on outdated electric bulbs. In most homes, old 100-watt incandescent bulbs, which consume excessive electricity, are still in use. Although the manufacturing of 100-watt bulbs has been discontinued, existing stocks are still sold and used.

Lighting in both domestic and public sectors predominantly depends on conventional, inefficient bulbs. LED bulbs offer superior light output and are 88% more energy-efficient than incandescent bulbs. Additionally, LED lights are 50% more efficient than CFLs.

On January 5, 2015, the Hon'ble Prime Minister of India launched the Unnat Jyoti by Affordable LEDs for All (UJALA) initiative to replace 770 million incandescent bulbs with LED bulbs by March 2019. According to the UJALA dashboard as of August 23, 2020, 36.63 crore LED bulbs have been distributed, saving 47.57 million kWh of energy, resulting in annual cost savings of Rs. 19,031 crore and reducing 3.85 crore tonnes of CO_2 per year. While there has been progress, the desired results have not been fully achieved. Many households continue to use traditional 100-watt bulbs and CFLs, highlighting the need for greater awareness and adoption of LED bulbs.

To reduce electricity consumption and safeguard the environment, it is essential to encourage the replacement of all non-LED bulbs in rural and urban households, offices, and businesses. By installing LED bulbs in every household, we can substantially reduce energy use, lower greenhouse gas emissions, and advance climate resilience.

Switching to LED bulbs is a step towards a sustainable and energy-efficient future!

31. Save Forests: The Power of Reducing, Reusing, and Recycling Paper

Despite the rise of digital communication, paper is one of the most widely used materials. From old newspapers to packaging materials and scrap printed sheets, the demand for paper continues to grow. Unfortunately, producing paper significantly impacts forests—24 fully grown trees are needed to make just one tonne.

According to the National Environment Bureau, recycling one tonne of paper can save nearly 13 trees, 2.5 barrels of oil, 4100 kWh of electricity, 4 cubic metres of landfill space, and 31,780 litres of water.

The solution to this problem is simple: adopting an attitude that prioritizes reducing paper consumption can help save forests and mitigate climate change.

The 3Rs - Reduce, Reuse, Recycle - offer practical ways to conserve paper and protect the environment:

Reduce:
- Minimize paper use by opting for digital documents whenever possible.
- Avoid printing unnecessary pages.
- Choose products with minimal packaging.

Reuse:
- Use both sides of the paper before recycling.

- Repurpose old paper, such as using them as notes or drafts.
- Share documents digitally to avoid multiple prints.

Recycle:
- Recycle paper products like newspapers, cardboard, and printer paper.
- Participate in community recycling programs.
- Support organizations that use recycled paper products.

By consistently applying the 3Rs, we can help conserve forests, reduce deforestation, decrease greenhouse gas emissions, and promote sustainable forestry practices to mitigate the effects of climate change.

Every small action contributes to a greener planet—adopt the 3Rs and positively impact the environment!

32. Reduce, Reuse, Recycle: A Powerful Step Towards Lowering Emissions

Practicing Reduce, Reuse, and Recycle (3Rs) can significantly reduce carbon emissions and mitigate climate change. Every product goes through a lifecycle--from manufacturing to disposal—and each step contributes to greenhouse gas emissions. By making conscious choices to reduce, reuse, and recycle, we can lessen our environmental impact and promote a more sustainable future.

Reduce:
- Purchase durable and long-lasting goods to minimize waste.
- Opt for products with minimal or no packaging, especially plastics, which can take over 200 years to decompose and release toxins.
- Decrease waste by choosing reusable items instead of disposable ones.

- Redesign products to use fewer raw materials, have a longer lifespan, or be easily reusable.

Reuse:
- Borrow items for short-term use rather than buying them.
- Donate or give away items usable to charity or individuals in need.
- Use reusable bags for shopping instead of plastic, which takes a long time to recycle.
- Opt for durable coffee mugs and refillable bottles.
- Reuse boxes, cloth napkins, and old magazines.
- Participate in paint collection programs and reuse containers for food storage.
- Refill pens and pencils when needed.

Recycle:
- Recycling prevents greenhouse gas emissions and reduces water pollution, saving energy.
- Using recycled materials generates less solid waste.
- Purchase products made from recycled materials rather than virgin materials.
- Recycle items such as paper, plastic, glass, and aluminum cans. Recycling half of household waste can save up to 1089 kg of carbon dioxide annually. Even a 10% reduction in household waste can save about 544 kg of carbon dioxide annually.

By consistently practicing the 3Rs, we can significantly decrease greenhouse gas emissions, conserve natural resources, reduce waste, and support sustainable consumption and production.

Every small action contributes to a healthier planet—adopt the 3Rs and make a lasting impact on the environment!

33. Embrace Sustainability: The Power of Buying Second-Hand

The increasing demand for new products contributes to environmental degradation and carbon emissions. Fast fashion and short-lived styles promote a throwaway culture, straining natural resources.

By choosing second-hand items—such as clothes, books, furniture, and electronics—you give products a second life, reduce waste, and conserve resources. Buying second-hand not only helps the environment but also lowers greenhouse gas emissions.

Second-hand items can be found in local stores or online platforms like OLX, Amazon, Flipkart, and eBay. Supporting second-hand purchases promotes sustainable consumption and contributes to a greener future.

Choose second-hand and make a positive impact on the environment!

Second-hand items can be found in local stores or on

34. Minimize Waste by Avoiding Idle Equipment

With rising incomes, many people buy items they only need occasionally. This leads to unnecessary spending and the waste of natural resources used in production. Instead, consider renting tools, equipment, or party supplies from specialty rental businesses or home improvement stores for occasional use.

Additionally, share items you already own with friends and family. Hosting block parties or encouraging a sharing culture in your neighborhood can foster a sense of community and environmental responsibility.

By avoiding idle equipment, we can reduce energy consumption, minimize waste, and promote sustainable practices, helping to mitigate climate change.

35. Donate Used Goods for a Sustainable Future

Donating used goods brings a sense of fulfillment and goodwill to individuals. As emotional beings, we often seek to make a positive impact. Instead of discarding used consumer items, donating them to those in need or charity centers is a meaningful alternative.

You'll feel better knowing that your used clothes, furniture, and home appliances will support someone who truly needs them rather than being stored away.

Donating used goods helps reduce waste, conserve resources, support sustainable consumption, decrease production demand, and minimize packaging waste.

Donating used items contributes significantly to a more sustainable future, reducing emissions, conserving resources, and helping mitigate climate change.

36. Choose Products with Minimal Packaging for a Greener Future

Excessive packaging, especially plastic-based packaging, contributes significantly to environmental waste. Packaging materials comprise over 30% of landfill waste globally, with around 207 million tons generated annually.

When disposed of improperly, excessive packaging drains natural resources, consumes energy, produces greenhouse gas emissions, and harms wildlife and ecosystems.

Many companies are adopting eco-friendly practices, such as using recyclable and renewable materials, reducing unnecessary packaging, and sourcing sustainably.

Choosing products with minimal packaging helps reduce waste, conserve resources, decrease emissions, and mitigate climate change. Every small change makes a big difference!

37. Switch to Reusable Containers for a Sustainable Lunch

Opting for reusable containers instead of paper bags for lunch can significantly reduce paper waste, lower greenhouse gas emissions, conserve resources, and minimize litter.

Reusable containers, such as stainless steel, glass, bamboo, and silicone bags, are durable, easy to clean, and environmentally friendly. They can decrease one's carbon footprint and inspire others to adopt sustainable practices.

Now switch to reusable containers today and contribute to a more sustainable future!

38. Choose Reusables Over Disposables for a Greener Future

The production and disposal of disposable products, such as paper and plastic plates, utensils, disposable diapers, and cheap plastic goods, contribute significantly to climate change. These items create substantial waste and remain in the environment for extended periods.

Avoiding disposable products helps reduce waste, conserve resources, lower emissions, decrease carbon footprint, and promote sustainable practices. Opt for reusable alternatives like cloth napkins, durable dishes, and refillable items to support a circular economy.

Make sustainable choices today!

39. Say No to Bottled Water: Embrace Reusables for a Cleaner Planet

Many opt for bottled water and cold drinks for convenience and comfort. However, globally, one million plastic bottles are bought every minute, and most of these end up in landfills or the ocean, taking over 500 years to decompose.

The production and transportation of plastic bottles consume significant energy and emit harmful greenhouse gases. Additionally, a small percentage of bottles are recycled, making the majority a long-lasting waste problem.

To reduce plastic waste and minimize environmental impact, opt for reusable water bottles or drink tap water through a glass.

By choosing reusable options, you support a sustainable future and help mitigate climate change!

40. Switch to Reusable Cups: A Simple Step for a Greener Tomorrow

Tea and coffee are the lifeblood of many people worldwide, helping keep us awake, productive, and energized. Starting the day with a cup of coffee or tea is a common habit for many.

However, when ordering takeaway drinks, most are served in disposable cardboard or plastic cups, contributing significantly to environmental waste. On average, drinking one cup daily can generate about 10kg of waste per year, harming the environment.

Reusable cups and mugs offer a simple, sustainable solution. Using them at work or in everyday life can significantly reduce waste. Additionally, encouraging tea stalls and cafes to adopt eco-friendly reusable cups is essential for promoting sustainability.

You're positively impacting the environment by choosing reusable cups and helping mitigate climate change!

41. Say No to Packaged and Frozen Foods: A Step Towards Sustainability

Ideally, the best way to ensure a healthy lifestyle is to buy fresh vegetables, groceries, meat, and fish from farmers' markets and cook at home.

Many people opt for frozen or pre-cooked foods available through various online platforms like Zomato, Swigged, Pizza Hut, Domino's, and McDonald's or directly from restaurants. These foods are often packaged in plastic containers or polythene, contributing to single-use plastic waste and harming health and the environment. The production of these packaging materials releases greenhouse gases, exacerbating climate change.

When dining out, consider the environmental impact and opt for fresh, locally sourced meals from nearby restaurants or food courts for a healthier, eco-friendly option.

Packaged Foods:

- Reducing packaging waste: Less plastic, paper, and cardboard waste.
- Conserves resources: Reducing packaging conserves materials.
- Lowering emissions: Manufacturing and transporting packaging generates emissions.
- Decreased food waste: Buying in bulk and planning meals reduces food waste.
- Supports sustainable agriculture: Whole, locally sourced foods promote sustainability.

Frozen Foods:

- Reduces energy consumption: It requires significant energy for freezing and lowering refrigeration.
- Lowering emissions: Transportation and storage generate emissions.
- Conserves water: Freezing foods uses water, so

reducing usage conserves resources.

- Decreases food waste: Cooking fresh foods reduces waste.
- Supports local food systems: Fresh, locally sourced foods promote local agriculture.

By avoiding packaged and frozen foods, you're making a positive impact by reducing waste, lowering emissions, conserving resources, and supporting sustainable practices, all of which contribute to mitigating climate change!

42. Say Goodbye to Single-Use Plastics: A Path to a Cleaner, Greener Future

We use single-use plastics daily, such as polythene bags, water bottles, tea/coffee cups, glasses, and containers for cold drinks/juices/oils/medicines. Without taking significant steps to reduce plastic pollution from production to consumption, it is predicted that by 2050, there could be more plastic in the oceans than fish.

The production, distribution, and disposal of single-use plastics require large amounts of fossil fuels, contributing to environmental pollution and climate change. Plastics persist in the environment, posing serious threats to ecosystems, marine life, and human well-being. Marine creatures ingest plastic; if we consume these sea creatures, we inadvertently consume plastic.

Avoiding single-use plastics can help:

- Reduce plastic production: Less demand for single-use plastics results in lower plastic production, which relies on fossil fuels and generates emissions.
- Decrease greenhouse gas emissions: The production of single-use plastics is energy-intensive, leading to higher emissions.

- Conserve resources: Using fewer plastics conserves non-renewable resources such as petroleum and natural gas.
- Minimize waste: Single-use plastics contribute to massive waste, and avoiding them helps reduce landfill waste.
- Reduce ocean pollution: Many single-use plastics end up in oceans, harming marine life and ecosystems.
- Support a circular economy: Avoiding single-use plastics promotes the circular economy, where resources are reused efficiently.
- Lower carbon footprint: Reusable alternatives reduce the carbon footprint associated with daily activities.
- Encourage sustainable practices: Avoiding single-use plastics raises awareness about plastic pollution and waste reduction.
- Save money: Reusable products can be cost-effective, benefiting both consumers and businesses.
- Set an example: Reducing single-use plastics inspires others to adopt sustainable practices.

By avoiding single-use plastics, you're moving towards a more sustainable future, reducing waste, emissions, and resource consumption, and contributing to the fight against climate change!

43. Curb Your Online Shopping to Minimize Environmental Impact

In recent decades, shopping has shifted dramatically from traditional street markets to online platforms where consumers can order products with a simple click and deliver them to their doorstep, often within a day. However, this trend comes with environmental consequences.

According to a World Economic Forum study, by

2030, urban last-mile delivery emissions are expected to rise by more than 30% in 100 cities globally. E-commerce sales have nearly tripled globally from 2014 to 2019, and with technological advancements, same-day and instant deliveries are becoming more prevalent.

This surge in online shopping contributes to increased emissions, pollution, and traffic congestion. High packaging materials, often consisting of single-use plastics, further exacerbate environmental harm. Additionally, as urbanization grows, so does the demand for faster home deliveries, leading to more delivery vehicles on city roads, adding to emissions and urban congestion.

While reducing online shopping entirely may be unrealistic, choosing stores that adopt eco-friendly practices and minimizing unnecessary purchases can help reduce environmental impact.

By curbing online shopping, you're contributing to a more sustainable future, lowering waste, emissions, and resource consumption, and aiding in climate change mitigation!

44. Choose Sustainable Fashion to Protect the Planet

The fashion industry is one of the most polluting sectors globally. Its environmental impact is significant, ranging from water usage to greenhouse gas emissions.

For example, producing a single pair of jeans requires 3,781 liters of water and emits approximately 33.4 kilograms of carbon dioxide. Additionally, the fashion industry contributes over 10% of annual global carbon emissions — more than all international flights and maritime shipping combined.

Fast fashion exacerbates these issues by promoting frequent design cycles and increased consumption. Each year, billions of garments are produced, with a substantial portion ending up in landfills or incinerated. Less than 1% of used clothing is recycled into new garments, resulting in significant value loss and environmental harm.

Choosing sustainable clothing, made from organic fabrics like cotton, hemp, and wool, can reduce these impacts. By supporting sustainable fashion, you're helping mitigate climate change, reducing waste, emissions, and conserving resources!

45. Recycling Clothes: A Simple Step for a Greener Future

Recycling clothes reduces greenhouse gas emissions by preventing them from decomposing in landfills, where they release harmful gases. Reusing fabrics conserves resources and reduces the energy and water used in manufacturing new clothing. Gently used clothes can re-enter markets with high demand for second-hand items, giving them a new life.

Landfills lack oxygen, causing organic materials like cotton to break down anaerobically, releasing harmful gases. Recycling just 100 million pounds of clothing is equivalent to removing 26,000–35,000 cars from the road annually.

By recycling your clothes, you help minimize waste, conserve resources, and contribute to a sustainable future, reducing your impact on climate change!

46. Proper E-Waste Disposal: Protecting the Environment and Conserving Resources

Electronic waste (e-waste) is a rising global issue, accounting for over 5% of municipal solid waste yearly. When burned or improperly discarded, items like phones, computers, and printers release harmful chemicals like PBDEs and greenhouse gases, posing risks to human health and the environment.

Instead of discarding broken electronics, consider repairing them or delaying upgrades for functioning devices. When it's time to dispose of electronics, ensure they are handled responsibly using certified e-waste recycling facilities. Always erase personal data from devices before disposal.

Recycling e-waste helps conserve valuable materials such as copper and aluminum, reduces pollution, and minimizes greenhouse gas emissions—up to 90% less than burning. Repair, reuse, or donate whenever possible, and avoid adding to plastic waste by disposing of electronics carelessly.

By adequately managing e-waste, you contribute to a cleaner environment, conserve resources, and help combat climate change!

47. Create Eco-Friendly Cleaning Products for a Greener Home

Switch to homemade cleaning products as a safer and eco-friendly alternative to commercial cleaners. They are gentler on the environment, animals, and people and can be stored in reusable containers, reducing plastic waste.

Did you know that over 50% of commercial cleaners contain harmful ingredients for the lungs, and about 20% include substances that can trigger asthma? Despite warning labels, many still use these products, which harm their health and the environment.

Creating cleaning supplies helps reduce chemical emissions, cut packaging waste, conserve resources, lower carbon footprint, prevent microplastic pollution, and reduce water contamination.

By making your own cleaning products, you support a sustainable lifestyle, minimize waste and pollution, and actively contribute to combating climate change!

48. Skip the Straw: A Small Change for a Big Impact

When we think of plastic pollution, we often picture bottles in rivers or marine animals entangled in plastic. Small, single-use plastic straws, used globally for convenient drinking, significantly contribute to this issue.

If everyone in Asia used a plastic straw daily, 4.5 billion straws would enter the waste system. Sustainable alternatives and increased awareness about reducing single-use plastics can make a big difference.

The next time you order a drink, say no to the straw. If you need one, opt for a reusable straw instead.

By skipping the straw, you help reduce plastic waste, decrease greenhouse gas emissions, conserve resources, protect marine life, and support initiatives for a plastic-free planet. Every small action counts toward a more sustainable and climate-friendly future!

49. Upcycling: A Creative Path to a Greener Future

While many are familiar with recycling, upcycling offers a creative and impactful way to protect the planet. Upcycling involves transforming existing materials into new, useful products, reducing the need for raw materials and minimizing energy usage, pollution, and greenhouse gas emissions.

Upcycling helps conserve valuable resources and reduces the environmental footprint of production and consumption by giving materials more than one life or a continuous cycle of uses. It also promotes creativity and sustainable living.

Upcycling contributes to waste reduction, resource conservation, pollution control, lower emissions, and energy savings while supporting sustainable practices and reducing climate impact.

You're actively fostering a healthier, cleaner planet and building a more sustainable future by upcycling!

50. Air-Dry Your Clothes for a Greener Tomorrow

In urban areas, many rely on electric dryers to dry clothes, which consume significant energy and contribute to greenhouse gas emissions. Opting for natural air drying instead can save energy and reduce one's carbon footprint.

Natural air-drying clothes is a simple yet impactful practice. By hanging your damp clothes on a line or rack and letting the sun or air do the work, you can reduce pollution and energy use. Green America says air-drying can reduce a household's carbon footprint by up to 2,400 pounds annually.

If you must use a dryer, choose an energy-efficient setting with automatic shut-off to minimize energy waste.

By drying your clothes naturally, you're helping build a sustainable future, lowering emissions, conserving energy, and fighting climate change!

51. Green Your Laundry: Eco-Friendly Tips to Reduce Your Carbon Footprint

Washing clothes can have a significant environmental impact, from water consumption to chemical pollution. Detergents, softeners, and bleach release harmful substances into water bodies, encouraging algae growth, depleting oxygen, and creating dead zones for marine life.

Here are simple tips to make your laundry eco-friendly and lower your carbon footprint:

- Invest in an efficient washing machine: Choose energy- and water-efficient models, such as front-loading machines that use only 60 liters per wash compared to 120–140 liters in other types.
- Wash full loads: Run the washing machine only when full to maximize efficiency. Avoid overloading, as it can stress the machine and reduce cleaning performance.
- Reduce washing frequency: Wash clothes only when necessary to save energy and water and extend the lifespan of garments. Some fabrics, like jeans and towels, can be used multiple times before washing.
- Use the proper cycles: Wash in cold water and opt for shorter cycles. Avoid "delicate" settings, which often use more water and release more microplastic.

By adopting greener laundry habits, you'll conserve resources, reduce pollution, and contribute to a sustainable future while combating climate change!

52. Reduce Tissue Paper Usage to Save Trees and Conserve Resources

Tissue paper is widely used daily, but its environmental impact is significant. Most tissue products are made from virgin pulp, which requires cutting down living trees, consuming vast amounts of water, and contributes to deforestation and pollution.

Environmental Impact of Tissue Paper

- Tree Consumption: Producing one ton of tissue paper requires at least 17 trees and contaminates 20,000 gallons of water.
- Water and Resource Waste: Worldwide, tissue paper consumption is immense, leading to deforestation and depletion of natural resources.

- Chemical Use: Soft tissue products are softer due to fresh wood pulp, increasing environmental harm.

Alternatives to Reduce Impact

- Washcloths: Using washcloths instead of single-use tissues can save countless trees and gallons of water.
- Handkerchiefs: Replace facial tissue with reusable handkerchiefs for a more eco-friendly option.
- Water Use: In many cultures, using water for personal hygiene can be more sustainable than relying on tissues.

Studies indicate small changes can significantly reduce tissue paper consumption and environmental impact.

By making eco-conscious choices, we contribute to a more sustainable future, reducing deforestation, emissions, waste, and pollution, and helping mitigate climate change!

53. Seed Balls: A Simple Solution for Promoting Greenery and Ecosystem Restoration

Seed balls, made from a mixture of mud, organic compost, and seeds, provide an easy and cost-effective way to promote greenery. They can be thrown into suitable land, even in difficult or inaccessible areas, allowing minimal maintenance and efficient growth.

In summer, collecting and preparing seed balls with local seeds can support plantation efforts during the monsoon season. Schools, colleges, and communities increasingly adopt seed ball initiatives to raise awareness and foster greener environments.

In 2019, I attended a seed ball camp at Utkal University, Odisha, organized by Mr Dharmendra Kar, a famous environmentalist and data scientist. Preparing seed balls is very easy, and I greatly enjoyed the speedball-making process. This is enjoyable for school/college students, communities, corporate employees, etc.

Despite a lower survival rate than saplings, seed balls offer a practical alternative for promoting green cover, especially in challenging terrains.

Supporting seed ball initiatives contributes to a sustainable future, restoring ecosystems and reducing environmental impact.

54. Seed Collection for Forest Creation: A Step Towards Sustainable Greening and Biodiversity

We consume various fruits daily, such as Mango, Lemon, Jackfruit, Papaya, Guava, Coconut, Bael, Orange, jamun, and Pomegranate. After these fruits are consumed, the seeds are often discarded into dustbins and eventually handed over to municipalities for disposal in garbage dumps.

Instead, you can collect, clean, and store these seeds from mature, non-rotten fruits. Seeds from Neem, Karanj, Arjun, Badam, Palm, and other native plants can be collected and stored in a dry place. During the monsoon season, these seeds can be dispersed in areas where planting saplings is challenging, such as forests, hillsides, or remote locations.

Collecting and storing seeds contributes to forest creation, biodiversity conservation, and climate change mitigation, fostering a more sustainable future.

55. Create Your Own Nursery: Empowering Green Spaces and Sustainable Growth

Everyone recognizes the importance of trees and their benefits. However, due to busy schedules, stress, and responsibilities, many find it challenging to plan and maintain tree plantations. Searching for a nursery for saplings further complicates the process, often resulting in the plan being abandoned.

Creating your nursery is a simple solution. You can grow saplings at home by collecting unused seeds from fruits consumed daily, such as mango, jackfruit, bael, and others. Use bottles, pots, or containers filled with sand and compost to nurture these seeds. Regular care will help them grow, and plant them in a safe location during the monsoon season.

This approach eliminates the need to depend on external nurseries and fosters a deeper connection with your plants. Creating your own nursery relieves stress and promotes health, environmental awareness, and community involvement.

Creating your own nursery contributes to reforestation, afforestation, and climate change mitigation while supporting biodiversity and community engagement!

56. Start a Kitchen Garden: Sustainable, Healthy, and Climate-Friendly

A Kitchen Garden is an excellent solution for combating climate change individually. Growing vegetable plants in your balcony, rooftop, lawn, or surrounding areas helps to absorb carbon dioxide, improves air quality, and provides fresh, organic produce.

Benefits include:

- Health Benefits: Fresh fruits and vegetables from your garden are nutrient-rich and free from harmful chemicals. Involving children in gardening encourages healthier eating habits.
- Cost Savings: Growing your own vegetables reduces the need to buy them from stores, lowering your monthly food expenses.

- Physical and Mental Well-Being: Gardening provides physical exercise and helps relieve stress, promoting a calm and rejuvenated mind.
- Natural Stress Reliever: Spending time outdoors in your garden improves mood and happiness.

Cultivating a kitchen garden contributes to sustainable agriculture, reduces emissions, enhances biodiversity, and supports climate resilience!

57. Adopt a Plant-Based Diet: A Powerful Tool for Climate and Health

Adopting a plant-rich diet is a significant solution for combating climate change. Dietary changes, though personal, can be influenced through education and awareness about the benefits of plant-based diets and the drawbacks of meat and dairy products.

Animal-based foods require more resources—water, land, and fuel—and contribute significantly to deforestation and biodiversity loss. In contrast, plant-based foods have a lower carbon footprint. For example, the carbon cost of beef is about 20 times more per gram of protein than beans.

A plant-rich diet offers numerous health benefits:
- Supports Immune System: Provides essential nutrients like vitamins, minerals, and antioxidants that strengthen the immune system.
- Reduces Inflammation: Neutralizes toxins and reduces the risk of inflammatory diseases like cancer and arthritis.
- Maintains a Healthy Weight: Removes foods contributing to weight gain, lowering the risk of related diseases.
- High in Fibre: Promotes better digestion, stabilizes

blood sugar, and lowers cholesterol, reducing cancer risk.

- Prevents Other Diseases: Reduces risks of heart disease, stroke, diabetes, and mental health issues.

By embracing a plant-rich diet, you support a sustainable food system, reduce emissions, and enhance climate resilience!

58. Choose Organic: Support Sustainable Agriculture for a Healthier Planet

The rising demand for food due to an increasing population has led to more intensive agriculture practices. However, using synthetic fertilizers and pesticides in conventional farming contributes to climate change and environmental degradation.

Synthetic pesticides and fertilizers are primarily derived from fossil fuels, which release greenhouse gases during manufacturing and transportation. Additionally, synthetic nitrogen fertilizers produce nitrous oxide—a greenhouse gas 300 times more potent than carbon dioxide.

In contrast, organic farming relies on natural methods such as manure and compost for fertilization. This practice enhances soil health, stores carbon, reduces soil erosion, and minimizes nitrate leaching into water sources. Organic farms are also more eco-friendly, with fewer pesticides and less energy usage. While organic farming often has lower yields, it compensates with numerous environmental and health benefits.

Benefits of organic food include:

- Fewer Pesticides: Organic produce contains lower levels of synthetic chemicals.
- Freshness: Organic food is often fresher, as it lacks preservatives.

- Environmentally Friendly: Reduces pollution, conserves water, and improves soil fertility.
- Animal Welfare: Organic livestock is raised without antibiotics or growth hormones.
- Nutritional Value: Organic meat and milk can prosper more in essential nutrients like omega-3 fatty acids.
- GMO-Free: Organic food is free from Genetically Modified Organisms (GMOs).

By choosing organic food, you promote sustainable agriculture, reduce emissions, and support a resilient environment!

59. Eat Local: Reduce Emissions and Support Your Community

Food choices are deeply personal, but few consider the environmental impact of how far their food travels. The distance food travels contributes significantly to carbon emissions, with transportation and packaging major contributors to greenhouse gas emissions.

Eating locally can significantly reduce these emissions, as locally sourced food requires less transportation.

Local foods offer numerous benefits beyond reducing transportation-related emissions:

- Full of Flavour: Locally grown crops are harvested at peak ripeness, ensuring fresh and flavourful produce.
- Seasonal Eating: Local foods are tied to seasonal availability, providing fresh and seasonal produce.
- Higher Nutritional Value: Shorter harvest-to-table times mean more nutrients in locally grown food.
- Supports Local Economy: Spending money on local food supports local farmers and reinvests in the community.

- Environmental Benefits: Locally grown food helps maintain farmland and reduces urban sprawl.
- Safer Food Supply: Local sourcing reduces contamination risks throughout the supply chain.
- Transparency: Local growers can provide details on how the food is grown and harvested.

Eating locally supports sustainable agriculture, reduces emissions, and fosters a stronger, more resilient community!

60. Reduce Consumption: Buy Less, Live More Sustainably

Daily objects — phones, clothes, furniture — carry hidden environmental costs. From production to transportation, these items contribute significantly to climate change.

Urban areas, particularly wealthy cities, have higher consumption rates, leading to increased carbon emissions. A simple T-shirt, for example, can have a complex supply chain involving energy-intensive manufacturing and shipping.

To reduce carbon footprints, here are some suggestions:

- Skip disposable packaging: Use refillable bottles and coffee cups.
- Embrace bartering and hand-me-downs: Swap items instead of buying new.
- Shop second-hand: Support thrift stores and online resale platforms.
- Fix what you already have: Mend clothing and repair appliances to extend their life.

Cities have started to play a crucial role in promoting sustainable living by implementing ambitious climate action plans. By consuming less and focusing on experiences over material goods, we can reduce emissions and lead a more sustainable lifestyle.

By buying less stuff, you're reducing emissions, conserving resources, and promoting sustainable consumption habits!

61. Bring Your Own Bag: Reduce Plastic Waste, Protect the Planet

Many people shop without carrying their own bags, leading to excessive use of single-use plastic bags. On average, each shopper uses 10 to 20 plastic bags for a single visit, which are then discarded into the environment.

Plastic bags contribute significantly to environmental harm:
- They take hundreds of years to decompose and break down into harmful micro plastics.
- Microplastics pollute oceans and waterways and harm marine life through ingestion and entanglement.
- The production of plastic bags consumes large amounts of energy and emits greenhouse gases.

By carrying a reusable bag, you can make a positive impact:
- Reduce plastic waste.
- Minimize greenhouse gas emissions.
- Promote sustainable consumption habits.
- Encourage friends and family to bring their own bags; collectively, we can create a cleaner, healthier environment.

Bringing your own bag reduces plastic waste, conserves resources, and promotes sustainable habits!

62. Minimize Food Waste: A Key to Sustainability and Climate Action

Approximately one-third of all food produced globally—about 1.3 billion tons—is wasted. This not only contributes to environmental harm but also exacerbates climate change.

Every time we waste food, we waste not just the food itself but also the resources that went into growing, processing, packaging, and transporting it. The rotting food in landfills produces methane, a greenhouse gas far more potent than carbon dioxide. By reducing food waste, we can significantly reduce greenhouse gas emissions.

Here are some tips to reduce food waste:

- Plan and buy what you need: Create a grocery list and avoid over-purchasing. Use leftovers first before buying new items.
- Use your freezer: Store leftover food to enjoy later. Frozen foods retain their nutritional value and last longer.
- Be creative with leftovers: Utilize overripe or imperfect produce to make smoothies, sauces, or soups.
- Create Awareness: Educate yourself and others about reducing food waste and its environmental impact.

By reducing food waste, you're conserving resources, reducing emissions and promoting sustainable food systems!

63. Home Composting: An Eco-Friendly Approach to Waste Reduction

Households generate waste, whether in urban or rural areas. Waste is typically disposed of in municipal bins in urban areas, while in rural areas, it's often left in open spaces for natural decomposition.

Waste disposal contributes to greenhouse gas emissions, especially when sent to open landfills, exacerbating climate change. Home composting is an effective solution to reduce these emissions.

Standard methods of home composting include:

- Aerobic Composting: Involves continual aeration of organic waste through air circulation and physical turning.
- Vermicomposting: Uses composting worms to assist in the decomposition process at ambient temperatures.
- Anaerobic Digestion: Takes place in oxygen-deprived environments with minimal temperature increase.

By practicing home composting, individuals can significantly reduce methane emissions and create nutrient-rich compost for sustainable gardening.

By composting at home, you're reducing emissions, creating nutrient-rich soil, and supporting sustainable agriculture!

64. Bio Enzymes: A Natural and Sustainable Cleaning Solution

Bioenzymes are organic cleaning solutions made through the fermentation of citrus fruits, jaggery, and water. They contain beneficial bacteria that produce enzymes to break down stains, soils, and malodors into smaller particles, which are then converted into carbon dioxide and water.

Benefits:
- Encourage sustainable living and are eco-friendly.
- Penetrates small crevices to eliminate stains effectively.
- Affordable compared to chemical cleaners.
- Support a zero-waste lifestyle.
- Do not pollute lakes or underground water.

Types of Enzymes:
- Proteases: Break down protein-based molecules like blood and food.

- Lipases: Break down fat and grease.
- Amylases: Break down starch molecules.
- Cellulases: Soften fabric and restore color.

Preparation:

- Combine jaggery, citrus peels, and water in a ratio of 1:3:10 with 1 tsp. yeast.
- Ferment in a plastic bottle for one month, stirring daily to release gases.
- Strain the liquid and store in a plastic container or bottles.

Application:

- Use as a multipurpose cleaner for surfaces like floors, glass, countertops, etc.
- Clean greasy utensils and repel insects like ants and cockroaches.
- Deodorise spaces with 1 part bio-enzyme solution to 3 parts water.
- Effective cleaning of drains and removing limescale and pesticides from fruits and vegetables.

By harnessing bioenzymes' power, we can reduce waste, emissions, and pollution while promoting sustainable practices and climate resilience!

65. Bring Your Own Container: A Simple Step to Reduce Plastic Waste

When dining out or taking food home, we often request that restaurant staff pack leftovers in plastic or polythene bags. These single-use plastics contribute to ocean pollution and increase carbon emissions during manufacturing and disposal.

While bringing your own containers for take-out may seem inconvenient, it is a necessary step for supporting the environment. Carrying your own container helps reduce plastic waste, conserve resources, lower greenhouse gas emissions, and promote sustainable practices.

Carrying your own container reduces plastic waste, conserves resources, and promotes sustainable consumption habits!

66. Opt for Ice Cream Cones: A Sweet Choice for the Environment

Many enjoy ice cream, often storing it in the freezer for later consumption. However, when eating ice cream outdoors, opting for a cone instead of a plastic cup is a more sustainable choice. This helps reduce plastic waste and prevents carbon emissions from manufacturing single-use plastic ice cream cups.

Choosing a cone supports environmental protection by conserving resources, saving energy, and promoting sustainable practices. Like choosing a cone, every small decision contributes to a more sustainable future.

Remember, every small choice counts, and opting for an ice cream cone is a sweet way to contribute to a more sustainable future!

67. Say No to Plastic Bottled Beverages: Choose Eco-Friendly Alternatives

Many enjoy sodas, juices, and other beverages for their taste and convenience. However, these are often packed in single-use plastic bottles, contributing to plastic waste and harming marine life when discarded. The production of these plastic bottles also emits carbon, contributing to environmental pollution.

Opting for homemade sodas and fresh fruit juices can provide health benefits while reducing reliance on sugary, packaged beverages. Choosing homemade alternatives or fresh fruits helps protect the environment by minimizing plastic waste and conserving resources.

By avoiding plastic-bottled beverages, you're reducing waste, conserving resources, and promoting sustainable practices!

68. Bring Your Own Container for Meat and Fish: A Sustainable Shopping Habit

A significant portion of the global population regularly consumes meat and fish. When visiting markets, many people pack these items in single-use plastics like polythene or plastic jars. After cooking at home, these single-use plastics are discarded in dustbins, contributing to environmental pollution. Eventually, these plastics may end up in oceans, harming marine life. Additionally, the production of these plastics results in carbon emissions.

Consider bringing your own container when purchasing meat, fish, or other non-vegetarian items to reduce this impact. This will minimize plastic waste, conserve resources, and help protect the environment.

Purchasing meat and fish in your own container can help reduce single-use plastics, conserve resources, and promote sustainable practices!

69. Buy Fresh Bread That Comes in Paper Bags: A Sustainable Choice to Reduce Plastic Waste

Breads are often packaged in plastic, sold in stores, or delivered through doorstep services. This plastic packaging contributes to pollution and carbon emissions during production.

Many companies are now shifting to paper bags or no packaging for bread. Choosing bread packed in paper bags or without any packaging can help reduce plastic waste and lower carbon emissions.

Buying fresh bread in paper bags or no bags reduces plastic waste, conserves resources, and promotes sustainable practices!

70. Choose Milk in Glass Bottles or Your Own Container: A Green Alternative to Plastic

Most people consume milk, which is often packaged in plastic or polythene. These milk packets or bottles are purchased from grocery stores, department stores, and other retailers. After use, they are discarded into dustbins, which are collected and stored at dockyards before eventually making their way to the oceans, harming sea habitats.

Many cities now have Milk ATM machines, allowing you to collect fresh milk directly without plastic packaging. Alternatively, you can purchase milk directly from farmers and use your own reusable container or glass bottles, an eco-friendly option.

Although convenience may lead to using packed milk, steps must be taken to reduce plastic waste and protect the environment.

Choosing milk in returnable glass bottles or your own container reduces plastic waste, conserves resources, and promotes sustainable practices!

71. Buy Loose Items: A Sustainable Way to Reduce Packaging Waste

Daily, you purchase staples like rice, dal, wheat, herbs, tea, and coffee from nearby grocery stores. Often, these items are heavily packaged in colourful plastic, with minimal focus on the product itself. After use, these plastic materials are discarded into dustbins, contributing to plastic pollution and eventually harming sea habitats when they reach oceans. Additionally, the production of these plastic packets emits carbon dioxide.

Why not carry your own container or reusable bags when purchasing loose items?

Many grocery stores now sell loose products like rice, dal, sugar, herbs, tea, and dry fruits, often at lower prices since packaging and labour costs are minimized. Loose items are often fresher and come directly from farmers' markets, unlike packaged goods stored in cold storage.

By purchasing loose items, you're reducing packaging waste, conserving resources, and promoting sustainable practices!

72. Carry Your Own Personal Care Products: Reduce Waste While Traveling

When you travel to different places and stay at hotels, you are often provided with free single-use personal care items like shampoos, soaps, lotions, toothbrushes, and more. These items, although convenient, contribute significantly to environmental damage as they are made from plastic and polythene. After use, they are discarded into dustbins by housekeeping staff, eventually reaching municipal waste facilities and, ultimately, the ocean, where they harm sea habitats. Additionally, the production of these personal care products emits a substantial amount of carbon dioxide.

Is it worth receiving these free items at the expense of the environment?

Consider carrying your own personal care items while traveling to minimize single-use plastics and protect the environment.

By carrying your own personal care products, you

reduce single-use plastics, conserve resources, and promote sustainable practices!

73. Save Water: Small Changes for a Sustainable Future

Water covers 71% of Earth's surface, but only 1% of the world's water is available as fresh, usable water for human consumption. Despite this limited resource, we use water for drinking, hygiene, cooking, and agriculture, putting significant strain on freshwater supplies.

Around 1 in 3 people live without safe drinking water, and global water demand is expected to increase by over 50% by 2040. Climate change exacerbates this issue through floods, droughts, and extreme weather events, making water more unpredictable and scarce.

Simple changes in our daily water use can help conserve this essential resource.

Here are a few tips to save water at home:

- Turn off the taps when not in use.
- Take shorter showers.
- Wash full loads of laundry.
- Install low-flush toilets.
- Water gardens early morning or late evening to avoid evaporation.

By saving water, you reduce energy consumption, conserve resources, and support sustainable practices!

74. Clean Your Village: A Collective Effort for a Greener Future

Approximately 65% of India's population resides in villages. In many villages, there is no municipal facility to maintain cleanliness. As a result, villagers take responsibility for cleaning their homes and immediate surroundings. However, surrounding areas, schools, and temple premises are often neglected and become polluted due to a lack of ownership. Garbage dumped in unused areas contributes to greenhouse gas emissions.

Our social responsibility is to keep our villages clean by placing garbage in designated composting areas.

Government schemes and funds are available through your local panchayat to support village cleanliness. You can utilize these by reaching out to your panchayat sarpanch or Block Development Officer.

Cleaning your village and surrounding areas promotes sustainability, reduces pollution, and supports climate resilience!

75. Save Water Bodies: Protecting Our Ecosystems for Future Generations

India is rich in traditional water bodies such as ponds, tanks, lakes, and talabs, which are crucial in maintaining ecological balance. These water bodies provide drinking water, recharge groundwater, control floods, support biodiversity, and offer livelihood opportunities.

India faces a severe water crisis, with over 100 million people affected by water scarcity. Projections indicate that by 2030, 40% of the population will lack access to clean drinking water.

In urban and rural areas, garbage, untreated sewage, and industrial effluents pollute water bodies. This pollution harms the environment and creates health issues for nearby communities.

Urban India is witnessing a rapid decline in the number and quality of water bodies. For instance, Bangalore has lost most of its lakes, and Ahmedabad has seen significant destruction of its lakes.

To save our water bodies, we must take action:

- Avoid polluting water bodies by not dumping garbage or puja items.
- Plant trees around water bodies to prevent soil erosion and maintain soil moisture.
- Preserve natural drains by preventing construction over them or dumping waste.
- Stop chemical pollution and inform local authorities of any violations.
- Treat industrial waste before discharging it into rivers.

Government schemes and funds are available to clean and conserve water bodies. To take advantage of these initiatives, you can collaborate with local panchayats and authorities.

By saving water bodies, you support ecosystem balance, regulate water cycles, and contribute to climate resilience!

76. Support Your Local Pond and River Clean-up: Preserve Ecosystems and Public Health

Thousands of ponds and rivers across rural and urban India are natural gifts that require our protection.

If you recall, there are likely local water bodies near your area where you've witnessed people throwing garbage, especially during religious ceremonies. This practice pollutes ponds and rivers with waste and releases harmful greenhouse gases.

Unfortunately, without regular clean-up efforts, these water bodies remain polluted. Municipal and panchayat authorities occasionally clean them when funding is available, but often, they are neglected.

Many people living near these water bodies suffer from waterborne health issues such as cholera, diarrhoea, and typhoid due to their reliance on polluted water sources.

You can make a difference by supporting local clean-up efforts. Even if government schemes aren't available, involving friends and neighbours in cleaning activities can significantly help maintain the health and ecosystem of your local ponds and rivers.

With your support, you're promoting ecosystem balance, reducing pollution, and contributing to climate resilience!

77. Plant Trees Around Water Bodies: Protect Water and Enhance Ecosystems

Ponds, lakes, rivers, reservoirs, and streams are vital in storing water in rural and urban India. These water bodies are essential for everyone, and we must protect them.

One key solution to safeguarding these water bodies is planting trees. Trees act as natural protectors by filtering pollutants, reducing heat, and enhancing water quality. They also help capture rainwater, preventing runoff and minimizing pollution that enters water bodies.

A study shows that a single tree can intercept an average of 1,685 gallons of rainwater each year, reducing excess runoff and conserving soil around water bodies.

Planting trees around water bodies supports ecosystem balance, reduces pollution, and contributes to climate resilience!

78. Green Your School: Inspiring Sustainability and Environmental Action

Schools are more than just institutions of learning; they are places where future leaders are nurtured and where environmental awareness begins. In today's world, combating climate change is essential for the survival of humanity and all living creatures. Schools play a crucial role in shaping responsible citizens who understand the importance of sustainable practices.

A Green School embodies environmentally friendly values, promoting a sustainable development model emphasizing a pollution-free, energy-saving, and eco-conscious environment. Schools can lead by example and inspire the community to adopt eco-friendly practices by setting environmental benchmarks.

Here are steps that students, parents, and teachers can take to transform a school into a Green School:

- Set up a green student club: Organize initiatives such as recycling programs, school gardens, clean-up drives, and green competitions.
- Cleanliness of Classrooms and School Surroundings.
- Implement Green Cleaning Policies: Ensure eco-friendly pest control and maintenance.
- Plant Trees: Beautify the school with greenery and oxygen plants in classrooms for fresh air.
- Encourage Sustainable Transportation: Promote walking, biking, carpooling, or using cleaner fuels like biodiesel in school buses.
- Plastic-Free Campaign: Advocate for reducing single-use plastics using reusable and recyclable materials.
- Energy Conservation Campaign: Educate students and staff on turning off unnecessary lights and computers and using energy-efficient devices.
- Install Solar Panels: Utilize renewable energy sources for electricity production and use solar lights on school campuses.
- Energy Meters: Display visible energy meters to encourage conservation.

- Green Schools Initiative: Encourage schools to sign sustainability and environmental responsibility initiatives.

By "greening" your school, you promote sustainability, reduce emissions, and inspire climate action!

79. Make Green Commuting Choices for School: Reduce Emissions and Promote Sustainability

In rural areas, students often travel to school by walking or cycling if the school is nearby. Sometimes, students use shared autos or are dropped off by parents on two-wheelers. However, in urban areas, many parents drop their children off by car or shared taxi for safety reasons, contributing to carbon emissions. Parents can consider more eco-friendly modes of transportation to reduce their environmental impact. Walking and cycling produce zero emissions.

Making green choices for school commutes can significantly help in combating climate change. Here are some eco-friendly options:
- Walking (0 emissions)
- Cycling (0 emissions)
- Biking (lower emissions)
- Carpooling (reduces emissions by 50-70%)
- Public Transportation (reduces emissions by 90-95%)
- Electric/Hybrid Vehicles (lower emissions)

By making green choices to go to school, you can help reduce emissions, improve air quality, promote physical activity, support sustainable education, and inspire others.

80. Family Planning: A Key Strategy for Sustainable Population Growth and Climate Action

The population of India was 16.90 Cr in the 1800s, grew to 23.84 Cr in the 1900s, and is currently 145 Cr (2024). Likewise, the World population was 100 Cr in the 1800s, rose to 160 Cr in the 1900s, and is presently 818.68 Cr (2024), growing at around 0.87% per year.

This rapid growth, driven by declining mortality rates due to health, sanitation, and infrastructure advancements, has increased resource demand, leading to higher greenhouse gas emissions and climate change.

A growing population required more food and infrastructure, which resulted in the emission of more greenhouse gases. Hence, family planning is required to curb population growth.

Family planning is one of the key adaptations for reducing greenhouse gases in the long run. We can't stop having sex for supporting climate change, but we can use contraception while having sex for avoiding unwanted pregnancies.

Many developing countries identify population growth as a challenge multiplier regarding climate change. Access to family planning reduces maternal and child mortality and produces better health outcomes, but it also strengthens climate change-affected communities' ability to adapt. With fewer unintended pregnancies, slower population growth reduces pressure on climate-sensitive resources.

During the COVID-19 Pandemic, millions of unwanted pregnancies happened around the world, which could have been prevented with the use of contraception.

There are over 89 million unplanned pregnancies annually worldwide, as 214 million women of reproductive age in developing regions want to avoid pregnancy but are not using a modern contraceptive method. If family planning were widely accessible and barriers to contraception were removed, unintended pregnancies, unplanned births, and abortions would decline substantially.

Having is the best possible individual climate action a person can take, but this is an individual choice and understanding. Better family planning is required, keeping the current climate crises in mind.

Family planning contributes significantly to healthier families, communities, and a healthier planet.

Supporting family planning empowers individuals, reduces emissions, and contributes to a more sustainable future.

81. Avoid Agricultural Biomass Burning: Protect Health and the Environment

With over 140 crore people (2024) and a major agro-based economy, India produces significant crop residues. Due to a lack of sustainable practices, many farmers burn this agricultural waste to clear fields, releasing harmful particulate matter that contributes to air pollution, global warming, and health issues.

Each autumn, Northern India is shrouded in smoke from burning 92 million tonnes of crop residues, creating toxic pollution visible from space. This practice emits black carbon, a potent climate pollutant that accelerates glacier melting in the Himalayas, endangering water resources for nearly 2 billion people. It also contributes to severe air pollution, causing millions of premature deaths annually.

Alternatives to open burning, like resource recycling, policy support, and farmer education, can help curb this practice. Initiatives by organizations like the Climate Change Coalition (CCC) and Clean Air Coalition offer solutions, such as subsidies for sustainable equipment and satellite fire monitoring.

Farmers must recognize the environmental and health hazards of biomass burning and adopt eco-friendly practices. Avoiding biomass burning reduces emissions, protects ecosystems, and ensures a healthier, more sustainable future for all!

82. Celebrate Diwali Responsibly: Avoid Firecrackers for a Greener, Healthier Future

The use of firecrackers during Diwali has been a cherished tradition for over a century. However, circumstances have changed significantly since then. In the early 1900s, India's population was around 23.84 crore; today, it has exceeded 140 crore. Despite this population explosion, the land area remains the same, leading to deforestation and increased strain on natural resources to meet the demands of food and infrastructure.

We must rethink our practices in light of the growing concerns about climate change and environmental pollution. Firecrackers contribute to air pollution, exacerbate climate change, and pose health risks, especially in today's densely populated and environmentally strained world.

While firecrackers created pollution even in the past, the current environmental challenges make it imperative to minimize further harm. Avoiding or opting for eco-friendly green crackers during Diwali can significantly reduce pollution levels. This small yet meaningful step will contribute to cleaner air and a healthier, more sustainable future.

Let's celebrate Diwali responsibly and work towards protecting our environment because every small effort counts!

83. Let's Ride a Bicycle: A Small Change for Big Environmental and Health Benefits

With the rapid acceleration of Science and technology and increased per capita income, humans have become physically lazy. They prefer to stay in their comfort zone and utilize their time effectively to generate income, rather than spending much time on their smartphones or social media. To effectively utilize time and avoid physical work, you can use a two-wheeler, car, auto, taxi, public transport, even for shorter commutes.

Commuting shorter distances emits significant carbon dioxide into the environment, contributing to climate change.

As per report on ET, Indian makers of passengers vehicles comprising cars, vans, and utility vehicles achieved corporate average fuel economy (CAFÉ)—a measure of carbon emission of a vehicle—of 116.078 gm per km in the year ended March 31 2023, missing the target of 113 gm per km set by the Ministry of Road Transport & Highways (MoRTH),

According to a 2015 report by the India GHG Program, the emissions for a Scooter with an Engine Capacity of less than 110cc are 36.8 g/km per ride, and a Scooter with an Engine Capacity of less than 150cc emits 38.7 g. A Motorcycle with an Engine Capacity of less than 100cc is 35.8 g/km per ride. Less than 125cc is 31.9 g, less than 135cc is 35.6 g, less than 200cc is 45.8 g, less than 300cc is 59.5 g, and less than 500cc is 59.7 g.

Daily, we commute shorter distances from one place to another within the city or village for office or business, purchasing rations, vegetables, visiting friends' and relatives' homes, etc.

Earlier, we commuted by Bicycle for shorter distances

within the city or village. Now, we use cars, motorcycles, or Scooters, which produce a lot of carbon dioxide.

Our offices, business places, ration shops, vegetable shops, and government offices are within 5 – 10 km of the city. If we commute 5–7 km to our office or business, we will reduce our carbon dioxide emissions. For Example, if your office is 7 km from your home, you have 14 km of travel for both sides. If you use a car, you emit 1,625.092 g CO_2 (116.078 g CO_2 emission per KM*14) to commute to the office. If you commute to your office for the same 7km distance via a Motorcycle of less than 135cc, then you emit 498.4 g CO_2.

If you commute 5 – 7km via Bicycle, you save 1,625.092 g CO_2 emissions in a Car and 498.4 g CO_2 emissions in a Motorcycle daily.

Bicycling is also a very good form of exercise from a health perspective. Now, everyone rides a Bicycle for fitness purposes. You can save on petrol expenses by cycling for shorter distances of up to 7km. Bicycling offers triple benefits: reduced CO_2 emissions, improved health, and savings on petrol expenses.

Now, let's have each individual have a Bicycle and use it for shorter-distance travel for daily activities. By using a Bicycle, you will make a significant contribution to environmental protection. Moreover, you will benefit significantly from having a bicycle at home.

Remember, every kilometre cycled instead of driven reduces emissions!

By cycling, you're reducing emissions, promoting sustainability, and contributing to a healthier environment!

84. Walking: A Small Step for You, A Big Leap for the Planet

The transport sector significantly contributes to

greenhouse gas emissions. We often use bikes or cars for short trips, such as buying groceries or commuting to nearby workplaces, even when the distance is less than 2 kilometers. These small trips result in unnecessary carbon emissions, which can be easily avoided through walking.

Walking to local shops or workplaces is a simple yet effective way to take climate action. Promoting a walking culture in our society benefits both the environment and individuals.

Beyond reducing emissions, walking enhances physical fitness and supports a healthier lifestyle. Every step you take contributes to sustainability, cleaner air, and improved well-being.

By choosing to walk, you're positively impacting the planet and your health!

85. Ride a Bike: A Greener Alternative to Cars for a Sustainable Future

With rising incomes, people are increasingly opting for luxury, and the purchasing power of Cars has increased. The Indian Auto sector is increasing rapidly. The total number of vehicles in fiscal year 2022 stood at 35.4 crore. Road travel was the preferred choice in India, with over 60 percent of the population using personal or shared vehicles for commuting.

In India, 34 people out of a thousand own a car. Most cars are located in urban areas. These cars are primarily used for commuting to work, purchasing groceries, and traveling from one place to another, which results in the emission of more greenhouse gases into the environment.

We can't stop communication, but we can reduce carbon emissions using a Bike. A bike can be used as an alternative to commuting to work, purchasing groceries, communicating, and doing other daily activities.

Bikes produce lower carbon emissions than cars. You can use a bike while keeping environmental protection in mind.

By riding a bike, you're contributing to a cleaner, healthier, and more sustainable future!

86.   Car Sharing: A Smarter, Greener Way to Travel

Car sharing is an effective way to reduce carbon emissions from vehicles. In India, several ride-sharing services, including Uber, Ola, Meru, Mega, and Carzonrent, offer shared ride options. Instead of using a personal car for single-use trips or purchasing a new car, individuals can utilize these services to share rides with friends or neighbors traveling to the same destination.

Car sharing reduces the need for car ownership, reducing carbon emissions and making transportation more cost-effective.

It's time to prioritize the environment by opting for shared rides instead of single-use car travel for work or other errands.

Car sharing is crucial in mitigating climate change, improving resource efficiency, promoting sustainable transportation, and reducing traffic congestion and emissions.

By embracing car sharing, you contribute to a more sustainable, efficient, and connected transportation system!

87.   Live Car-Free: Embrace a Healthier, Greener, and Sustainable Lifestyle

Living car-free is a crucial climate action that supports a healthier and more sustainable urban environment. Cars have revolutionized travel, offering convenience and speed, but their dependence has significant environmental and health consequences. Cities, designed to accommodate vehicles, have reshaped streets to prioritize cars, often at the expense of pedestrians and public spaces.

The widespread use of cars contributes to over 70% of global CO_2 emissions from the transport sector, which reduces air quality and leads to traffic-related injuries and chronic health issues. Urban areas, especially in metro cities, experience increased pollution levels due to the rise in cars, worsening year by year. For example, Delhi has faced severe air pollution, prompting government efforts to reduce carbon emissions, which can only be achieved with individual action.

Cities worldwide are shifting towards reducing car dominance to create healthier environments and improve quality of life.

A car-free lifestyle contributes significantly to a cleaner, healthier, and more sustainable future!

88. Avoid Air Travel: A Simple Choice for a Greener, More Sustainable Future

Flying contributes considerably to carbon emissions into the atmosphere for those who take regular holidays abroad or domestically and travel for business or personal reasons.

Around 2.5% of global CO2 emissions come from aviation. Together with other gases and the water vapor trails produced by aircraft, the industry is responsible for around 4% of global warming.

At first glance, that might not seem like a very big contribution. Except, only a tiny percentage of the world flies frequently. Even in richer countries like the UK and the

US, around half of people fly in any given year, and just 12-15% are frequent fliers.

In India, the percentage of flying is very low compared to rich countries. Still, we need to think and can reduce carbon emissions by avoiding unnecessary travel.

The COVID-19 pandemic taught us how to manage business by doing video conference calls and avoiding travel. Business is happening as usual without domestic or international travel. We can prevent domestic or international holidays and enjoy our lives by staying home, strengthening our family values.

Before preferring air travel, you may ask yourself a few questions by keeping environmental protection in mind.

Is this business trip necessary?

Can it be managed via tele-video conference?

Can it manage via Rail travel?

Can it be managed via any other mode of travel?

Is this Holiday necessary?

Can this Holiday be managed by visiting domestic places instead of international places?

Can we avoid this Holiday by staying at Home to improve family values?

Now is the time to consider the environment before going on an air trip. Also, create awareness and advise your colleagues, friends, and relatives on alternatives to air travel. In this way, you can contribute to the environment.

Avoiding air travel reduces emissions, promotes sustainability, and creates a cleaner environment!

89. Avoid Unnecessary Travel: Protect the Planet and Embrace Sustainable Living

People often travel for vacations, holidays, business, or personal trips using air, rail, cars, or bikes. While these travels provide joy and togetherness, they often overlook the environmental impact. Each journey, whether by air, rail, car, or bike, contributes to carbon emissions by burning fossil fuels.

Unnecessary travel not only incurs financial costs but also causes long-term environmental harm. Reducing such travel can help combat climate change by lowering greenhouse gas emissions, minimizing transportation-related pollution, promoting virtual alternatives, supporting local economies, and encouraging sustainable transportation practices.

Eliminating unnecessary travel helps reduce emissions, supports sustainability, and fosters a cleaner, healthier environment!

90. Avoid Cruise Holidays: Choose Sustainable Travel and Protect Our Oceans

As people's incomes rise, there is an increasing preference for more entertainment and unique vacation destinations. While traditional vacations involve air, train, bus, or car travel, a new trend has emerged with Cruise Holidays. Cruise vacations occur on ships that sail across oceans, stopping periodically at ports for sightseeing. Popular cruise destinations include Bombay, Goa, Kerala, Kochi, Chennai, the Andaman Islands, and Odisha. The cruise market is rapidly expanding as more people are drawn to this form of travel. The Tourism Ministry predicts that the cruise tourism market could reach ₹35,500 crore by 2041, serving around 4 million passengers.

By avoiding cruise holidays, you contribute to climate change mitigation, promote sustainable tourism, protect marine ecosystems, reduce air and water pollution, and encourage eco-friendly travel choices.

91. **Switch to Electric Cars: Drive Cleaner, Greener, and More Sustainable**

Electric cars produce fewer emissions than conventional vehicles powered by gasoline, diesel, or gas. Emissions are categorized into Direct and Life Cycle emissions.

- Direct emissions occur from the tailpipe, fuel evaporation, and refueling processes. All-electric vehicles produce zero direct emissions, improving air quality in urban areas. Plug-in hybrid electric vehicles (PHEVs) have a gasoline engine and electric motor, producing fewer emissions than conventional vehicles, though they may have some evaporative and tailpipe emissions.

- Life cycle emissions include all fuel and vehicle production, distribution, use, and disposal emissions. EVs generally have lower life cycle emissions than gasoline or diesel due to cleaner electricity generation. Using renewable energy sources like solar and wind can further reduce these emissions.

Although EVs are slightly more expensive than traditional vehicles, government incentives, such as cash subsidies, reduced road tax, and free registration, are increasing to promote their adoption for a carbon-free India.

Consider switching to an electric vehicle for your next car purchase to reduce carbon emissions and support a sustainable transportation future!

92. **Choose Electric Two-Wheelers: Clean, Green, and Efficient Transportation**

Electric two-wheelers are set to be a game-changer in reducing carbon emissions. The electric two-wheeler market in India is rapidly growing, with more people opting to purchase EVs for their next ride.

Choosing an electric two-wheeler for short-distance travel is a step towards a carbon-free India.

Using electric two-wheelers can lead to zero emissions, lower carbon footprints, enhanced energy efficiency, reduced air pollution, support for sustainable transportation, and decreased reliance on fossil fuels.

By choosing an electric two-wheeler, you're helping to create a cleaner, healthier, and more sustainable transportation system!

93. Choose Electric Rickshaws: Clean, Green, and Sustainable Urban Transport

Electric rickshaws are an environmentally friendly and clean mode of transport compared to conventional auto rickshaws run on gasoline (petrol, Diesel, and Gas). They reduce air pollution on city roads. Electricity is required to charge batteries, and carbon emissions may occur during electricity production. However, electricity can be generated from renewable sources like solar and wind rather than burning coal.

Currently, in India, e-Rickshaws can be seen in Railway stations, Bus stands, city roads, and rural areas due to numerous Government subsidy schemes.

Now we must choose E-Rickshaw rather than conventional rickshaw, keeping environmental pollution in mind as responsible citizens.

Using an electric rickshaw contributes to a cleaner, healthier, and more sustainable transportation sector!

94. Go Electric: Replace Your Vehicle for a Cleaner, Greener Future

With rising incomes, everyone has money and can pay to live a luxurious life and stay healthy. But environmental degradation and clean air can't be bought with money. Only small individual actions can upgrade our environment and clean our air.

Replacing your existing car and two-wheeler with an electric car and two-wheeler is one way to protect your environment. Electric Cars and Two-Wheelers can save carbon emissions from your existing conventional vehicles run by gasoline (petrol, diesel, and gas). Electric cars and two-wheelers emit zero carbon.

Replacing your current car and two-wheeler with an electric vehicle (EV) can significantly help mitigate climate change by reducing emissions, lowering your carbon footprint, and improving energy efficiency.

Replacing your current car and two-wheeler with an electric vehicle significantly contributes to a cleaner, healthier, and more sustainable transportation sector!

95. Switch to LPG: Cleaner Cooking for a Healthier and Greener Home

The number of active domestic LPG consumers has grown significantly, from 14.52 crore in April 2014 to 31.36 crore as of March 2023. This rise is primarily attributed to the Pradhan Mantri Ujjwala Yojana (PMUY), which increased LPG coverage from 62% in 2016 to 104.1% in 2022. As of January 30, 2023, 9.58 crore connections have been provided under PMUY. Additionally, on March 24, 2023, the Union Cabinet approved a subsidy of Rs . 200 per 14.2 kg cylinder for up to 12 refills per year for PMUY beneficiaries.

Launched in May 2016, PMUY aims to improve women's and children's health by providing free LPG connections to underprivileged families. Initially focused on families below the poverty line, the scheme has since been extended to families above the poverty line and other eligible categories.

The Indian government has made significant efforts to combat household air pollution, ensuring LPG access for every household through the Ujjwala Yojana.

It is now the responsibility of each household to use LPG gas instead of solid fuels like biomass and firewood to protect the environment.

Using LPG for household cooking creates a cleaner, healthier, more sustainable energy future!

96. Solar Cooking: A Sustainable Solution to Reduce Carbon Emissions

Solar cooking uses solar thermal energy to cook food without relying on electricity. This eco-friendly method uses a renewable energy source that does not pollute the environment.

India, with its extensive agricultural population, benefits significantly from solar energy. It receives 5-7 kWh per square meter of solar energy for about 300-330 days a year. Solar cookers reduce the use of polluting fuels like coal, firewood, and LPG, minimizing emissions and the risk of respiratory diseases.

The Ministry of New & Renewable Energy (MNRE) actively supports promoting and distributing energy-efficient solar cookers in India.

While solar cooking may not be feasible for everyone, it is particularly beneficial in rural areas where solar energy is abundant. It effectively reduces carbon emissions compared to traditional cooking methods like chullas or LPG.

By adopting solar cooking, you contribute to a cleaner, healthier, and more sustainable energy future!

97. Say No to Open Defecation: Protect the Environment and Promote Sanitation

Open defecation pollutes the environment. Open defecation is the human practice of defecating outside ("in the open") rather than into a toilet. People choose open fields, bushes, forests, ditches, streets, canals, or other spaces for defecation. They do so either because they do not have a toilet readily accessible or due to traditional cultural practices. The practice is common where sanitation infrastructure and services are not available. Even if toilets are available, behaviour change efforts may still be needed to promote the use of toilets. 'Open defecation free' (ODF) is a term used to describe communities that have shifted to using toilets instead of open defecation. This can happen, for example, after community-led total sanitation programs have been implemented.

Under the Swachh Bharat Mission (SBM) scheme, over 10.9 crore toilets have been built across India, and as of 2 October 2019, rural India was declared free of Open defecation. However, not all households use their toilets for defecation due to multiple issues, like a lack of sanitisation infrastructure. The government has done its duty, but more awareness is required for 100% toilet usage.

Now, awareness needs to be created among people, and people need to think about the environmental damage caused by open defecation and do their part as responsible citizens for environmental protection.

Using toilets instead of open defecation contributes to a healthier, more sustainable, and climate-resilient world!

98. Eco-Friendly Festivals: Celebrate with Nature in Mind

Festivals in India bring joy and unity to people

across the nation. However, these celebrations often come with environmental consequences. Activities like bursting firecrackers during Diwali, excessive water usage during Holi, and immersing idols in water bodies contribute to pollution and harm the environment.

To celebrate festivals in a more eco-friendly manner, we can adopt several practices:

Ganesh Chaturthi, Durga Puja, Viswakarma Puja, etc.:

- Use organic and biodegradable materials like clay and mud for idols.
- Opt for local immersion tanks to minimize damage to water bodies.
- Choose natural colors like turmeric or geru for painting idols.
- Encourage eco-friendly competitions using sustainable materials.

Diwali:

- Focus on light, not noise. Go noise-free.
- Avoid leaving decorative lights on unnecessarily.
- Choose eco-friendly crackers with lower decibel limits or alternatives like sparklers.
- Use natural and recycled materials for rangolis and decorations.

Holi:

- Use herbal gulal or household items to create colors like turmeric, sandalwood, or henna.
- Practice dry Holi to conserve water.
- Avoid burning wood for Holika bonfires; use recycled materials instead.

By celebrating festivals in an eco-friendly way, we contribute to a more sustainable future and help mitigate climate change. Every small change counts, and collective efforts can have significant positive impacts!

99. Support Ecotourism: Preserve Nature and Empower Local Communities

Ecotourism plays a crucial role in conserving the environment while promoting the well-being of local communities. It involves the participation of forest fringe dwellers and those living away from forests, ensuring a balanced interaction between nature and human activity.

Developing ecotourism in wildlife conservation areas and other natural landscapes helps preserve ecosystems while providing livelihood opportunities for local communities. Responsible ecotourism focuses on preserving natural features and promoting sustainability.

By supporting ecotourism, you contribute to climate change mitigation and foster sustainable development!

100. Choose Ecolabel Products: Make Sustainable Choices for a Greener Future

Ecolabel products are designed to have less environmental impact than traditional products. They provide information about the environmental standards met during their production process, helping consumers make informed choices.

In India, the EcoMark scheme, launched by the Government in 1991, ensures that ecolabel products meet high environmental and quality criteria. Various products, including batteries, cosmetics, food items, and textiles, have incorporated these standards.

By choosing ecolabel products, you support sustainable practices, reduce your carbon footprint, and contribute to climate change mitigation!

101. Buy Sustainably Harvested Wood Furniture: Protect Forests and the Planet

Furniture plays a significant role in our homes and offices. However, the wood used for furniture often comes from unsustainable sources, leading to environmental damage such as ecosystem disruption, loss of wildlife, and waterway pollution.

Opting for furniture made from sustainably harvested wood ensures that forests are managed responsibly, preserving them for future generations. Sustainable forestry helps absorb carbon dioxide, maintain air quality, and support biodiversity.

By choosing sustainably harvested wood furniture, you support responsible environmental practices and climate change mitigation!

102. Save Energy and Reduce Emissions by Turning Off Your Computer and Laptops

Many people use computers and laptops at home or work, which consume significant energy. These devices are often left on throughout the day, even when not in use. Additionally, leaving them on for extended periods contributes to carbon dioxide emissions, primarily from coal-powered electricity.

A screensaver is not an energy-saving solution. According to the U.S. Department of Energy, 75% of electricity is used to power electronics in standby mode. An average desktop computer consumes 60 to 250 watts daily, and leaving it on continuously results in higher energy use and carbon emissions. By turning off your computer when not in use, you can save up to $70 annually and significantly reduce CO_2 emissions.

Here are some tips to save energy and reduce carbon emissions:
- Turn off your computer at night or when not in use.
- Enable Power Management features for your monitor.

- Turn off the monitor when it is idle for 15 minutes or more.
- Consider laptops over desktops, as they consume 1/4 the energy.
- Opt for flat-screen monitors, which use 1/3 the energy.

Turning off your computers and laptops can help reduce energy waste, lower emissions, and promote a more sustainable future!

103. Switch Off Lights: Save Energy and Protect the Environment

Most of the time, lights are on in rooms and offices when unused. This is a careless issue or an awareness issue. You may not pay electricity bills at your home or office for free, but you put the environment at stake because of your careless approach. Lighting bulbs consume electricity, electricity is produced by burning fossil fuel, and fossil fuel burning means the emission of greenhouse gases. Unnecessary light produces greenhouse gases. The emission of greenhouse gases means increasing air pollution, which affects numerous health issues.

Please turn off your lights when you don't need them, whether in your home, office, school, college, clubhouse, public place, hotel room, etc. This will also reduce your electricity bills and air pollution.

Turning off lights when unnecessary reduces energy waste and emissions and contributes to a more sustainable future!

104. Boost Energy Efficiency: Use Power Strips in Your Home Office and Entertainment Centre

We use energy in our daily lives; we can't live without it, and we rely on it for everything. We need electricity to light rooms, power computers and laptops, charge mobile phones, run electronic/electrical appliances, keep our entertainment centre live, etc.

We know that energy is consumed when we use it, but we are unaware that these household items also drain power when switched off. Consuming electricity means burning more fossil fuel, which results in the emission of greenhouse gases.

Power strips can play a big role here. Installing a smart power strip in your home is a quick and easy way to save money while making your household more energy efficient. This small amount of energy savings can significantly save the environment.

Using power strips reduces energy waste, lowers emissions, and contributes to a more sustainable future!

105. Unplug Devices When Not in Use to Save Energy and Protect the Environment

We generally switch off electronic devices when they aren't in use, and think there is no energy consumption after switching them off. However, we are unaware that devices drain energy when they are off. So, these energy drains during device switching off cost the environment as electricity is produced by burning fossil fuels.

Unplugging devices like computers, laptops, TVs, Mobile phones, Home appliances, etc., can save electricity. The saving percentage is not high, but a little energy savings by everyone can significantly help the environment.

Unplugging electronic devices when not in use reduces energy waste, lowers emissions, and contributes to a more sustainable future!

106. Reducing Energy Consumption with Efficient Devices

Energy is part of our lives. Every day, we rely on it and consume electricity. Electricity produced by burning fossil fuels like coal emits greenhouse gases that affect climate change. We can't stop consuming energy, but we can reduce our consumption by installing energy-efficient devices in our homes and offices.

Home appliances, including clothes, washers, dryers, dishwashers, refrigerators, freezers, air purifiers, and humidifiers, will account for approximately 20 percent of your home's total electric bill. Energy-efficient devices can reduce the percentage of energy consumption. Lower energy consumption reduces electricity bills, protects the environment by reducing the emission of greenhouse gases, and enhances your lifestyle. Also, these energy-efficient devices benefit your country and the world by protecting the environment.

Examples of energy-efficient devices include
- LED bulbs
- Energy Star-rated appliances
- Smart thermostats
- Power-efficient computers and laptops
- Energy-efficient Water heaters

Using energy-efficient devices reduces energy waste, lowers emissions, and contributes to a more sustainable future!

107. Brightening Cities Sustainably with LED Lights

Every city requires lights for lighting. Most cities have had traditional lights for a long time, consuming more electricity and producing more greenhouse gases. Greenhouse gases contribute to climate change.

Cities can save energy by illuminating public spaces with LED lights and replacing existing traditional lights with LED lights. LED lights are more energy-efficient than traditional lights. Tell your Municipality authorities and Panchayat Sarpanch for replacing existing traditional lights to LED lights for protecting environment.

By switching to LED bulbs, cities can:
- Reduce energy consumption by up to 90%
- Lower greenhouse gas emissions
- Save money on energy costs
- Enhance public safety and security
- Promote sustainable development

You're contributing to a more energy-efficient, sustainable, and climate-resilient future by lighting up your city with LED bulbs!

108. Switch to E-Bills and Online Payments for a Greener Future

Traditionally, utility companies like electricity, telephone, mobile, and water generate physical bills and send them to consumers via postal or courier services. Consumers often visit utility offices to pay their bills in person, using cheques or cash. This process consumes paper, contributing to deforestation and waste accumulation at municipal dockyards.

Additionally, traveling to utility offices for bill payments involves burning fossil fuels, resulting in greenhouse gas emissions contributing to climate change.

Nowadays, most companies offer E-Bill facilities, allowing consumers to view and manage bills online by entering their account details. Online bill payment is also available, eliminating the need for physical bill printing.

Opting for e-bills and online payments can help save millions of trees and reduce greenhouse gas emissions.

By preferring e-bills and paying bills online, you reduce paper waste, conserve resources, and support a more sustainable future!

109. Create an Eco-Friendly Home with Green Building Plans

A home is a sanctuary where we seek comfort, peace, and prosperity. Residential energy use contributes approximately 4% of greenhouse gas emissions, enhancing the environmental impact of our homes.
How can we build our dream homes while minimizing energy consumption to reduce carbon emissions?

Design your home using advanced technology to control heat, air, and moisture leakage by sealing doors and windows. Insulate your garage and basement with natural, nontoxic materials such as reclaimed blue jeans. To maximize energy efficiency, shield your windows from sunlight with large overhangs and double-pane glass. Prioritize natural cross-ventilation and install solar panels on your rooftop to harness solar energy, reducing dependence on traditional sources like coal. Integrating solar panels and other energy-saving modifications for existing homes can significantly reduce energy consumption.

By getting blueprints for a greenhouse, you'll promote sustainable living, reduce energy use, and support a more eco-friendly future!

110. Conduct an Energy Audit for a More Efficient Home

In the face of rising environmental challenges and the detrimental effects of climate change, everyone is responsible for minimizing greenhouse gas emissions at the individual level. Your home is the first step you can take. To effectively reduce your energy consumption, it is essential to understand how much energy your home uses and identify ways to reduce it. You can conduct an energy audit yourself or enlist the help of a professional. While energy-saving technologies have evolved rapidly and are commonly available to builders and architects, a professional energy audit accurately assesses your home's energy consumption.

During an energy audit, you can locate and seal air leaks; inspect ventilation; assess insulation; evaluate lighting, appliances, and electronics; and inspect heating and cooling systems.

Once you identify areas where your home is losing energy, create a plan by considering the following questions:

- How much do you spend on energy?
- Where are your most significant energy losses?
- How long will it take for an investment in energy efficiency to pay for itself through energy savings?
- Do the energy-saving measures offer additional benefits, such as increased comfort from efficient windows?
- How long do you plan to stay in your current home?
- Can you handle the work yourself, or do you need a contractor?
- What is your budget?
- How much time can you allocate for maintenance and repairs?

By conducting an energy audit, you reduce energy waste and emissions and contribute to a more sustainable future!

111. Opt for Energy Star Appliances to Save Energy and the Environment

When purchasing a car, you often compare fuel mileage to make an informed decision. Why not do the same when buying home appliances from supermarkets or stores?

You regularly purchase home appliances such as Air Conditioners, Refrigerators, TVs, Coolers, Microwaves, Grinders, Dryers, Iron Boxes, Washing Machines, Induction Cookers, Pressure Cookers, Kitchen Utensils, Home Theatres, and heating/cooling devices. While standard home appliances may be cheaper, Energy Star appliances are designed to be more energy-efficient.

Many companies now manufacture Energy Star appliances and offer them to consumers. Despite potentially higher upfront costs, Energy Star appliances can save you money in the long run by reducing your electricity bills. Additionally, lower energy consumption results in fewer greenhouse gas emissions, supporting a healthier environment.

When choosing home appliances, consider their long-term benefits and environmental impact.

Buying Energy Star appliances reduces energy waste and emissions and contributes to a more sustainable future!

112. Keep Your Tires Properly Inflated for Better Fuel Efficiency

Many of you rely on cars or bikes for daily commuting. Due to busy schedules, checking your tires regularly may be overlooked. Tires with low pressure lead to decreased mileage and increased fuel consumption—higher fuel consumption results in more greenhouse gas emissions, negatively impacting the environment.

Regularly check your tires and ensure they are properly inflated to reduce fuel consumption and protect the environment.

Keeping your tires properly inflated reduces emissions, conserves energy, and contributes to a more sustainable future!

113. Turn Off Your Vehicle Engine to Save Fuel and Reduce Emissions

We commute daily to work, home, and other destinations using our cars or bikes. Often, we encounter traffic or pause for conversations with friends, forgetting to turn off the engine while waiting. Additionally, we may answer phone calls while parked by the roadside, leaving the engine running. Many believe that idling consumes less fuel than driving, but this is not true. Keeping the engine on unnecessarily increases fuel consumption and greenhouse gas emissions.

Turning off your vehicle engine when not in use—at traffic lights or during short stops—can significantly reduce fuel consumption and lower carbon emissions. This simple habit protects the environment.

Turning off your vehicle engine reduces emissions, conserves fuel, and contributes to a more sustainable future!

114. Wash Your Car at a Service Station for Eco-Friendly Cleaning

Everyone wants to fit themselves and have regular health check-ups at Path Labs. When seeking medicines, we prefer to take them after consulting with Doctors rather than directly from Medicine stores.

Likewise, everyone wants their car to be in good working condition and neat and clean for optimal performance. We regularly wash it at a service station or at home. When we wash our car at home, we use more water but cannot wash it properly like at a service station, where the car's performance is optimal. This results in more fuel consumption and water loss, meaning more carbon emissions.

A service station is the right place for efficient and optimal car performance. Experts can wash your car with less water.

To make washing your car more eco-friendly, consider the following:

- Wash your car only when necessary.
- Choose a service station with eco-friendly practices and equipment.
- Use eco-friendly cleaning products at home.
- Consider using a waterless car wash or a car wash with a rainwater harvesting system.

Remember, every small action counts, and making conscious choices can help mitigate climate change!

115. Upgrade Your Air Conditioner and Refrigerator for Energy Efficiency

Most households use air conditioners and refrigerators to cool their homes and stay comfortable during the summer. However, these appliances consume significant electricity, leading to higher energy bills. Air conditioners and refrigerators use more energy than other household appliances, including light bulbs. Increased energy use produces higher greenhouse gas emissions, contributing to climate change and its severe impacts.

To reduce your electricity bill and environmental impact, consider upgrading your old air conditioner and refrigerator to newer models with advanced, energy-efficient technologies. Although upgrading may require an initial investment, it will save you money on monthly electricity bills while reducing energy consumption. This not only benefits your wallet but also helps protect the environment.

Upgrading your air conditioner and refrigerator will reduce energy waste, lower emissions, and contribute to a more sustainable future!

116. Reduce Energy Waste by Emptying and Closing Your Refrigerator Before Traveling

We travel from one place to another for our requirements, so we lock our home if other members are not traveling to stay at home. We often forget to switch off our refrigerator, as only a few foods or vegetables are in it. We may empty our refrigerator and switch it off before leaving home for a long time to save energy.

- Pre-Travel Checklist: Consume or dispose of perishable items, Remove all contents, Clean and dry the interior, Unplug the refrigerator, Set the temperature to "off" or "vacation mode.", and Leave the doors slightly ajar for airflow.
- Additional Tips: Plan meals before traveling, Store non-perishable items, Donate unspoiled food, Check manufacturer guidelines, and unplug other appliances.
- Travel Duration Guidelines: Short trips (<1 week): Unplug refrigerator; Medium trips (1-4 weeks): Empty and unplug; Long trips (>4 weeks): Clean and store refrigerator.

By emptying and closing your refrigerator before traveling, you'll reduce energy consumption, minimize food waste, and extend the lifespan of your appliance.

Make it a habit before your next trip!

117. Save Energy and Improve Air Quality by Maintaining Your HVAC Filters

HVAC filters, such as the ones for your air conditioner, play a vital role in your home's heating and cooling system. They trap dust and other small particles, ensuring the air you and your family breathe is cleaner. By filtering these particles, the filters prevent them from being recirculated throughout your home.

If you've noticed increased dust around the house, it could be due to a dirty HVAC air filter. As the weather cools and you spend more time indoors, an old filter might contribute to extra dusting and higher energy consumption, negatively impacting the environment.

To optimize performance, it's recommended that you clean your HVAC air filter every 2-3 months and replace it every 3-6 months. This benefits the environment by conserving energy, reducing electricity costs, improving indoor air quality, and promoting better health for your family.

Keeping your HVAC air filters clean or replacing them regularly can reduce energy waste and emissions and help create a more sustainable future!

118. Save Water and Energy by Installing an Automatic Water Level Controller

Water is life; without water, we can't live. Water is a gift from god, and it shouldn't be wasted. In most houses, a water tank is available. We fill the water tank with a motor pump using manual operating switches. After switching the motor, we wait 15 -30 minutes to fill the water tank. When the water overflows from the tank, we switch off the motor and use this water for the entire day. Sometimes, we cannot predict whether water will be available and face difficulty when going for a bath or toilet. Hence, we always switch off our motor after the water overflows from the tank. In this way, more electricity is consumed and water is wasted. More electricity consumption means more emissions of greenhouse gases, which causes climate change. Also, water wastage is not good for the environment as only 1% of clean water is available on the earth for drinking.

Installing an automatic water level controller in your tank prevents water wastage and reduces electricity consumption. This device regulates the water level and automatically switches the motor off when the tank is full, ensuring no manual intervention is needed. It maintains optimal water levels throughout the day, saving water and electricity.

Using an Automatic Water Level Controller conserves precious resources, promotes energy efficiency, and contributes to a sustainable future!

119. Advocate for Green Energy and Clean Electricity in Your Community

India's primary energy source is coal, with most electricity generated from coal-based plants. However, the country is steadily progressing toward its goal of generating 50% of its electricity from renewable sources by 2030.

You can contribute to this transition by encouraging your utility companies to purchase and supply clean energy to your area. Renewable electricity, derived from solar, wind, and hydropower sources, is more affordable than traditional electricity generated from coal and gas and significantly less harmful to the environment. While coal-based energy contributes to pollution, renewable energy is clean, sustainable, and eco-friendly.

By choosing Green Energy and advocating for the

adoption of clean electricity, you help drive the shift to a low-carbon economy and contribute to a greener, more sustainable future!

120. Design Your Workspace to Maximize Natural Light and Save Energy

We spend 40% of our day at our workplace. Generally, workplaces are in multi-story buildings lit with electricity and have an air conditioner for office hours. The Lighting Bulb and air conditioner consume electricity produced from burning coal. Coal burning produces greenhouse gases, which cause climate change.

We can design our office surrounded by natural light, which helps us save electricity. Light bulbs may not be required during the daytime. Saving electricity can help save the environment, while also lowering electricity bills. Natural light can boost productivity, engage workers, and enhance their well-being.

By incorporating natural light into your workspace, you can reduce energy consumption, promote sustainable design, and contribute to a more environmentally friendly future.

121. Implement Sustainable Practices to Create an Eco-Friendly Workplace

Climate change affects working individuals, including those experiencing heat exhaustion, heatstroke, infectious and pollution-related diseases, malnutrition, injuries from extreme weather events, and reduced productivity. Making small, thoughtful changes to your workplace can significantly contribute to environmental protection while enhancing your work environment.

- Add Indoor Plants: Place a few plants on your desk

and around the office. These not only purify the air and absorb carbon dioxide but also enhance productivity, alleviate stress, and foster a positive atmosphere.

- Switch to Energy-Efficient Lighting: Replace traditional lights with LED bulbs to save electricity and enjoy brighter, more natural lighting.
- Minimize Printing: Avoid printing unless absolutely necessary. Opt for digital reading via emails and messaging apps to save paper and protect trees.
- Upgrade to Energy-Efficient Electronics: Invest in newer, energy-efficient air conditioners and other devices to reduce electricity consumption.

By making these changes, you can reduce energy use, promote sustainable practices, and contribute to a greener, more climate-conscious future.

122. Grow Your Own Food and Support Community Gardens for a Sustainable Future

Food is essential for life, but how it is sourced often comes with hidden environmental costs. Many foods we buy travel hundreds of kilometres to reach our grocery stores, impacting their freshness and flavour while generating significant greenhouse gas emissions during transportation. Additionally, the packaging materials used, such as plastics and cardboard, create waste and require energy-intensive manufacturing processes that also emit greenhouse gases.

Commercially produced foods are often grown using chemical fertilizers, the specifics of which are often unknown to consumers, and are stored in energy-consuming cold storage facilities, further contributing to greenhouse gas emissions. Protecting the environment and waterways from pollution and

Growing your own food can break free from the traditional food supply chain. This reduces the need for long-distance transportation and packaging, reducing fossil fuel use and waste. Supporting community gardens is another way to promote sustainable practices while fostering a sense of connection and shared purpose.

When you grow your own food, you control what goes into the soil and the fertilizers used, ensuring your produce is free from harmful chemicals. This practice protects the environment and waterways from pollution, giving you peace of mind and satisfaction in knowing exactly what you eat.

By cultivating your own garden or contributing to a community garden, you're actively reducing emissions, supporting sustainable agriculture, and helping to build a healthier, more climate-conscious future.

123. Choose Fuel-Efficient Vehicles for Lower Emissions and Cost Savings

Cars and bikes play a vital role in our daily lives, helping us commute from one place to another. When purchasing a new vehicle, prioritize models with fuel-efficient designs and low carbon emissions. These vehicles reduce your fuel costs and help protect the environment.

Consider opting for efficient and less polluting options, such as electric vehicles, that align with your needs. Avoid using older vehicles that contribute significantly to pollution, keeping environmental preservation in mind.

By choosing fuel-efficient vehicles, you're reducing emissions, conserving energy, and actively contributing to a more sustainable future!

124. Report Electricity Theft to Help Conserve Energy and Protect the Environment

Electricity theft is a significant issue in India, occurring in rural and urban areas. Despite achieving village electrification across almost all regions, the problem persists. Individuals involved in electricity theft often use power irresponsibly, treating it as a free resource. They may bypass electricity meters to run appliances like electric heaters and air conditioners or collaborate with inspection officials to lower their bills. This misuse leads to unnecessary electricity consumption, increased greenhouse gas emissions, and contributes to climate change.

In contrast, honest consumers carefully monitor their electricity usage to manage costs, while electricity theft users exploit the system without considering the financial or environmental impact.

By reporting electricity theft to your local electricity office, you can help save energy and promote fairness. When a complaint is filed, the electricity department will conduct an inspection, and penalties will be imposed on those found guilty. This also acts as a deterrent for others considering theft.

By reporting electricity theft, you are helping to reduce energy waste, conserve resources, and contribute to a more sustainable and equitable energy system.

125. Report Environmental Violations to Protect Our Natural Resources

Under the Constitution of India, the environment is a fundamental right encompassed within the Right to Life under Article 21. Every individual has the right to a clean and healthy environment. Article 51-A of the Constitution also places a duty on the State to protect and improve the environment. Natural resources, including air, water, land, flora, and fauna, must be preserved for the benefit of both present and future generations through effective planning and management.

Despite numerous environmental laws in India, violations by individuals, corporations, and government agencies continue infringing on others' rights. Common environmental violations include:

- Unauthorized tree or forest cutting.
- Improper waste disposal.
- Illegal discharge of pollutants into water bodies.
- Use of banned pesticides in agriculture.
- Contamination of drinking water sources.
- Emission of pollutants like particulates, sulfur, nitrogen, and carbon beyond regulatory limits.
- Oil spills, wetland destruction, and open burning.
- Non-compliance with environmental guidelines and permissions.

Citizens can lodge complaints about such violations with authorities such as the local Forest Office, Municipal Corporation, Police Station, Central Pollution Control Board (CPCB), State Pollution Control Boards, or the National Green Tribunal (NGT), depending on the nature of the case.

Individuals can file complaints or initiate legal cases against individuals, corporations, or government bodies for failing to protect the environment, violating environmental laws, or causing harm due to environmental negligence.

Only a few environmental violation cases are reported in India. To raise awareness and promote accountability, it is crucial to increase the frequency of such complaints.

By filing complaints, citizens encourage adherence to environmental laws and foster greater environmental consciousness.

Complaining about an environmental violation is helpful, but only a few cases are reported in India. To raise awareness and promote accountability, it is crucial to increase the frequency of such complaints, protect the environment, hold violators accountable, and contribute to a sustainable future. Let us act now to safeguard our planet for future generations.

126. Support or Start an Urban Farm for Sustainable Food and Reduced Emissions

In urban areas, we often rely on grocery stores for fruits and vegetables, which typically travel hundreds to thousands of kilometres to reach us. This mode of transportation requires burning fossil fuels, while packaging materials generate additional waste. Both activities contribute significantly to greenhouse gas emissions and climate change.

Urban farming, which involves growing food in and around cities, offers a sustainable alternative. It utilizes available urban spaces efficiently and can produce more food per area than traditional farming methods. If you don't have time to farm, you can still support friends or community initiatives engaged in urban agriculture. Urban farming reduces the need for long-distance food transport and helps absorb carbon dioxide from the urban environment, improving air quality.

Starting or supporting an urban farm helps reduce emissions, encourages sustainable agriculture, and contributes to a healthier, more climate-friendly future.

127. Work from Home One Day a Week to Reduce Emissions and Support Sustainability

The COVID-19 pandemic highlighted the importance of health and environmental sustainability. Lockdowns and restrictions kept us at home, prompting individuals and organizations to explore ways to work remotely. Many corporate and government organizations have adapted by implementing remote work systems, providing necessary tools such as laptops and secure system access. This shift revealed that working from home often increases productivity, as employees can focus better in a quiet, familiar environment.

The work-from-home culture has been widely adopted globally. By working remotely, individuals avoid commuting, which reduces fossil fuel consumption, greenhouse gas emissions, and traffic congestion. This benefits the environment and saves valuable time and travel expenses. During lockdowns, cities experienced significantly lower pollution levels due to reduced vehicle usage.

Working from home at least one day a week can help maintain these environmental benefits. It's a simple yet impactful way to contribute to reducing emissions and supporting a cleaner, greener planet.

By working from home one day a week, you can reduce emissions, conserve resources, and contribute to building a more sustainable future.

128. Embrace Work-from-Home Jobs to Reduce Emissions and Promote Sustainability

The concept of Working from Home (WFH) gained prominence during the COVID-19 lockdowns, which restricted people from commuting or traveling. Companies quickly adapted by developing processes and products to support remote work, equipping employees with necessary IT infrastructure and system access. Since 2020, WFH has become a widely accepted practice among organizations and employees.

Both employees and employers benefit from WFH. Employees appreciate flexible work environments, reduced commuting time, and the ability to work from anywhere, resulting in a better work-life balance and increased productivity. On the other hand, employers save on infrastructure costs such as office space, utilities, and operational expenses.

Environmental Benefits of WFH:

- Reduced commuting emissions: Up to 75% less CO_2 emissions from transportation.
- Lower energy consumption: Approximately 30% less energy is used compared to office setups.
- Decreased transportation infrastructure demands: Less need for roads and public transport.
- Minimized urban sprawl: Reduced land use for office spaces and related developments.
- Less paper usage and waste: Digital workflows decrease reliance on physical documents.

Additional Benefits:

- Improved work-life balance: Flexible schedules allow employees to manage personal and professional responsibilities.
- Increased productivity: Employees often perform better in a comfortable, distraction-free environment.
- Reduced traffic congestion: Fewer commuters result in less crowded roads and public transportation.
- Enhanced health and well-being: Lower stress levels and better time management.
- Cost savings: Employees save on commuting costs, and employers save on operational expenses.

Discuss the possibility of remote work with your employer, if your job role allows it. By embracing a work-from-home (WFH) approach, you reduce your carbon footprint, promote sustainable living, and support efforts to mitigate climate change.

Let's make WFH a climate-friendly norm and work toward a more sustainable future!

129. Choose a Work-from-Home Job to Support Sustainability and Improve Work-Life Balance

The concept of Working from Home (WFH) emerged during the COVID-19 pandemic and has continued to thrive post-pandemic. Due to its mutual benefits, it is now widely embraced by employers and employees. Many companies offer WFH options, and employees increasingly prefer these roles for the comfort, convenience, and work-life balance they provide, along with freedom from traffic and daily commutes.

Opting for a WFH job also aligns with environmental sustainability. By reducing the need for commuting, WFH jobs help lower greenhouse gas emissions, decrease energy consumption, and minimize urban congestion.

Benefits of Choosing a WFH Role:

- Environmental Impact: Reduced carbon footprint and fossil fuel usage.
- Improved Quality of Life: Achieving a better work-life balance and a stress-free work environment.
- Global Sustainability: Contributing to efforts to combat climate change.
- Cost and Time Savings: No commuting means saving on travel expenses and reclaiming valuable time.

By joining a company that supports WFH, you'll improve your personal and professional life and make a meaningful contribution to the environment and the planet's future.

Take the first step toward a greener, climate-friendly career today!

130. Track and Measure Your Building's Green Performance for a Sustainable Future

Just as we monitor our progress and achievements in areas like education, health, or financial planning, it is equally important to track the environmental performance of our buildings. Climate change effects—such as rising global temperatures, extreme heat waves, droughts, floods, and cyclones—make prioritizing sustainability and minimizing environmental impact essential.

Tracking your building's performance in key areas, such as energy use, water consumption, waste management, transportation, and occupant experience, is vital. Regular monitoring helps identify opportunities for improvement and ensures alignment with environmental goals. Digital platforms, designed with user-friendly interfaces and aligned with green building rating systems, can facilitate this process.

One platform, Arc, allows you to measure and compare your building's performance while connecting metrics to green building strategies. Adopting such tools can improve your building's performance, lower energy consumption, and reduce greenhouse gas emissions, thereby supporting environmental conservation.

By tracking and measuring your building's green performance, you contribute to increased energy efficiency, reduced waste, and a more sustainable future for everyone.

131. Monitor Local Air Pollution and Take Action for Cleaner Air

Monitoring air quality is crucial for understanding

the levels of cleanliness or pollution in the air, which directly impact both health and the environment. Air quality is typically assessed using the Air Quality Index (AQI), which measures the concentrations of pollutants such as PM2.5 and PM10 — microscopic particles that can impact respiratory health. The AQI ranges from 0 to 500:

AIR QUALITY INDEX

- AQI under 50 indicates good air quality.
- AQI from 50 to 100 is moderate.
- AQI from 100 to 150 is unhealthy for sensitive groups.
- AQI from 151 to 200 is unhealthy for everyone.
- AQI from 201 to 300 is very unhealthy.
- AQI from 301 to 500 is hazardous.

You can measure local air quality using available devices or online resources. If you find that your local air quality is poor, take action by raising awareness and promoting measures to reduce air pollution. This can include reducing fossil fuel use, avoiding open biomass burning, using clean cooking stoves, and minimizing diesel emissions from transportation.

Benefits of Acting to Reduce Air Pollution:

- Identifies Pollution Sources: Helps pinpoint sources for targeted reduction efforts.
- Reduces Greenhouse Gas Emissions: It lowers emissions from fossil fuels and industrial processes.
- Improves Public Health: Reduces respiratory and other pollution-related health issues.
- Promotes Sustainable Transportation: Encourages eco-friendly transportation options, such as walking, cycling, or electric vehicles.

- Enhances Energy Efficiency: Optimizes industrial practices to reduce energy consumption.
- Enhances Quality of Life: Improves overall community well-being and health.
- Inspires Others: Sets an example and encourages community-wide action.
- Informs Policy: Provides data to support policies and regulations aimed at achieving cleaner air.

By measuring and taking action to reduce local air pollution, you're contributing to improved public health, lower emissions, and a more sustainable future.

132. Advocate for 100% Clean Energy in Your City

As a developing nation, India is the world's third-largest energy consumer and a significant contributor to greenhouse gas emissions. While its per capita carbon emissions remain lower than the global average, India has made significant commitments under the Paris Agreement, including a goal to generate 50% of its energy from renewable sources by 2030. The country is making notable progress in renewable energy, particularly solar energy production.

Cities transitioning to renewable energy sources experience reduced pollution and contribute to a cleaner, healthier environment. As a responsible citizen, you can advocate for 100% clean energy in your city by raising awareness, engaging with local representatives, and encouraging the administration to prioritize renewable energy initiatives. While achieving 100% renewable energy in India may not be immediately feasible due to the country's reliance on non-renewable sources, your efforts can help foster change and promote the adoption of eco-friendly energy solutions.

By advocating for 100% clean energy in your city, you're helping drive the transition to a sustainable, equitable, and climate-friendly future!

133. Practice Digital Hygiene: Clean Up Your Inbox and Reduce Energy Consumption

In this digital world, we can't imagine our lives without email. We receive emails daily from different companies as advertisements or communication.

Did you know that every old email stored in your inbox consumes energy? Decluttering your inbox is a quick and easy way to reduce your electricity consumption and shrink your carbon. We often overlook the fact that these junk mails are impacting our environment.

Generally, e-mails appear to save resources. Unlike traditional letters, no paper or stamps are needed; nothing has to be packaged or transported. Many assume that email requires little more than electricity to power our computers. It's easy to overlook the invisible energy usage involved in running the network and maintaining the entire physical infrastructure behind it, particularly when sending and storing data.

Every email in every inbox worldwide is stored on a server. The incredible quantity of data requires huge server farms—gigantic centres with millions of computers that store and transmit information. These servers consume massive amounts of energy 24 hours a day and require a considerable amount or air conditioning systems for cooling. The more messages we send, receive, and store, the more servers are needed, which means more energy is consumed and carbon emissions are produced.

Why It Matters:

- Spam emails release an estimated 0.3 grams of CO2

each, while emails with attachments can emit up to 50 grams.

- Junk emails account for over half of all emails globally, with significant potential to reduce emissions through better inbox management.
- Deleting unnecessary emails can save server energy and reduce water usage for cooling massive data centres.

Steps to Declutter Your Inbox:

- Delete Old Emails: Regularly clean your inbox, especially those with attachments.
- Unsubscribe: Opt out of unnecessary newsletters and spam.
- Turn off Notifications: Disable redundant email alerts from social media platforms.
- Use Green Email Providers: Opt for eco-friendly email services that are powered by renewable energy.

Impact of Junk Mail Reduction:

- Saves millions of trees and gallons of water and prevents substantial carbon emissions.
- Reduces paper waste in landfills and lessens identity theft risks.
- Frees up your time while contributing to a sustainable digital ecosystem.

By eliminating junk mail, you're reducing your digital carbon footprint and helping to promote a more eco-friendly future.

134. Invest in Renewable Energy and Divest from Fossil Fuel for a Sustainable Future

As the impacts of climate change intensify, it's imperative to oppose polluting energy sources and champion clean, renewable alternatives. One of the most effective actions individuals and organizations can take is to divest from fossil fuels and reinvest those funds into initiatives that benefit both people and the planet. Divestment involves withdrawing financial support from fossil fuel companies, such as stocks and bonds, recognizing these investments as harmful to the environment and increasingly risky for investors.

Fossil fuel divestment is vital in reducing carbon emissions by accelerating the adoption of renewable energy sources. It also exerts public pressure on companies engaged in fossil fuel extraction to transition toward greener alternatives. This movement has gained significant momentum, with over $14.58 trillion worth of investments divested globally by governments, corporations, educational institutions, religious organizations, and pension funds.

Investing in renewable energy companies is wise for those who want to protect the environment and support sustainable growth. Renewable energy fosters innovation, creates jobs, and mitigates climate change, making it a pivotal sector for the future.

Directing your investments toward renewable energy and avoiding fossil fuels drives a sustainable energy transition, contributing to a cleaner, greener, and more climate-resilient world.

135. Invest in Innovations for Climate Change Solutions to Accelerate Sustainability

Generally, we invest money in the market to generate returns in various sectors, including education, services, manufacturing, and banking. However, as the effects of climate change intensify, human society's survival is at risk. The Earth's temperature has increased by 1 degree Celsius above pre-industrial levels due to the rising emissions of greenhouse gases. The world, including India, is investing substantially in renewable energy and climate change solution innovation companies to address this crisis.

Climate change solution innovation companies require substantial funding for research and development. By investing in these companies, you provide the necessary resources to accelerate innovation and create sustainable solutions. While the returns on these investments may be lower, they play a crucial role in combating climate change and protecting the planet.

Following the Paris Agreement in 2016, some of the world's wealthiest individuals, including Bill Gates, established funds focused on technology-driven climate change solutions. These initiatives bring together governments, research institutions, and billionaire investors to tackle climate change. Their investments primarily support research in new technologies for power generation, transportation, food production, manufacturing, and building sectors, all aimed at reducing greenhouse gas emissions.

By investing in climate change solution innovations, you contribute to the development of groundbreaking technologies, the scaling of effective solutions, and the fostering of a more sustainable future.

136. Educate Your Children to Become Climate-Conscious and Advocates for a Sustainable Future

Climate change is a complex and important topic that can be challenging to explain to children. However, it is essential to empower them to understand what climate change entails and how they can make a difference.

Parents are crucial as the first teachers in a child's life. How we guide and teach our children shapes their perception of the world. Often, we focus on control—

enforcing discipline, adhering to strict schedules, and making decisions on their behalf. But true learning comes from nurturing, allowing children to understand and experience the world on their own terms.

The same principle applies to nature. We often try to control nature for our benefit rather than nurturing and caring for it. Children need to feel connected to the environment to instil a deeper understanding and appreciation of climate change. They need to understand that preserving nature stems from love and care.

Provide your child with opportunities to engage with nature, such as planting trees or gardening, and explain how these activities contribute to reducing carbon emissions and protecting the planet. Begin with simple activities and gradually introduce concepts such as reducing waste, conserving water, and promoting sustainability.

By inspiring children to fall in love with nature, you encourage them to act responsibly. Educating them about climate change through meaningful experiences will help shape a generation aware, engaged, and committed to creating a sustainable future.

Through this approach, you are fostering responsible leadership and cultivating a climate-conscious mind-set in your children, ensuring they become advocates for the planet.

137. Empower Girls Through Education to Foster a Climate-Conscious and Resilient Future

Educating girls has been shown to have a significant impact on climate change. Research highlights that educated women are better equipped to understand the dangers of global warming and are more prepared for disasters. They are better equipped to protect their families and contribute to community resilience.

According to UNESCO, educating girls could substantially reduce 51.48 gigatons of emissions by 2050. This is because educated girls have a positive influence on their immediate families and communities, leading to beneficial effects. Studies have consistently shown that women with higher levels of education tend to marry later, have fewer and healthier children, live longer, and experience greater economic prosperity.

For instance, in Mali, women with secondary education or higher have an average of three children, compared to seven children for those with no education. Additionally, women's increased income has a more significant impact on their children's nutrition and health than men's.

Investing in girls' education also reduces population growth. The United Nations projects that with the continued expansion of girls' education, the global population could be reduced by two billion people by 2045. Furthermore, educating girls enhances human development in health, economic progress, and democratic governance.

Despite progress, around 130 million girls worldwide still lack access to quality education. This disparity is due to factors like poverty, community traditions favouring boys, and limited resources for educational development. Women's historical disadvantages, including restricted access to resources and decision-making, make them particularly vulnerable to climate change impacts.

In developing countries, women heavily depend on local natural resources for their livelihoods and often face limited mobility. They are often responsible for securing water, food, and fuel, which exposes them to more significant climate risks.

Empowering girls through education fosters

a generation of leaders who will drive sustainable development and create a resilient, climate-conscious future.

138. Join Climate Education to Become an Advocate for Change and a Solution-Oriented Leader

Over the past 78 years since India's independence, the population has grown by over 100 crore, and the impacts of climate change are becoming increasingly severe, manifesting as rising temperatures, extreme heatwaves, floods, cyclones, and high pollution levels in cities both in India and worldwide. To address these challenges, climate education is essential.

While awareness about climate change is growing, it is still progressing at a slow pace. Fortunately, numerous online platforms offer climate education resources. By participating in climate education, you can gain a deeper understanding of the causes, impacts, challenges, and potential solutions to climate change. Climate education helps foster a sense of responsibility towards the environment and empowers individuals to take action to protect the planet through their personal efforts.

By learning about climate solutions, you can share this knowledge with your family, friends, and the community, inspiring collective action and contributing to a more informed and proactive society.

Joining climate education initiatives empowers you to enhance climate literacy, inspire meaningful action, and equip future leaders to combat climate change effectively!

139. Take Part in Environmental Awareness Rallies or Protests to Amplify Your Voice for Climate Action

With the rising impacts of climate change, numerous

social groups, NGOs, and environmental advocates are organizing awareness camps and rallies at various levels — villages, cities, districts, state levels, and beyond — with the mission of environmental protection. These initiatives include planting and saving trees, reducing plastic and air pollution, addressing water pollution, promoting family planning, supporting girls' education, and participating in the Clean India Mission.

You can actively participate in these environmental awareness rallies and protests to show your support. Sharing your involvement on social media helps inspire and encourage others to join in. Your participation can inspire friends, family, and community members to take collective action, ultimately impacting the environment.

By supporting environmental awareness rallies, you're contributing to your own environmental well-being and helping to create a sustainable and healthier planet. These efforts bring a sense of fulfilment in making a positive environmental difference.

Engaging in environmental awareness rallies or protests can help you advocate for climate action, inspire meaningful change, and secure a sustainable future!

140. Support or Join Youth-Led Movements for Climate Change to Empower the Next Generation of Environmental Leaders

Environmental enthusiasts, social workers, and students from schools and colleges are leading movements to raise public awareness and protect the environment from the impacts of climate change. These movements, often driven by young people, aim to bring about positive change for the betterment of society. However, many of these initiatives face challenges due to limited financial resources and insufficient public support, which hinders their ability to sustain momentum.

You can play a vital role in these youth-led movements by providing financial and moral support. Many young individuals, such as students and environmental advocates, often face unemployment and rely on their families and friends for support. Your contribution can help keep these movements alive and drive forward the cause of environmental protection.

Additionally, your support can amplify these efforts through social media, spreading awareness and encouraging others to join in.

Supporting or joining youth-led movements empowers people to drive climate action, innovation, and sustainable change!

141. Join Organizations Working for Environmental Protection to Contribute to Collective Climate Action and a Sustainable Future

Climate change is causing a continuous rise in the Earth's temperature, leading to more frequent extreme weather events, such as heatwaves, cyclones, floods, ecosystem disruptions, and rising sea levels. The Indian government is actively working to allocate funds toward climate change protection through various schemes, with a focus on renewable energy production. The Ministry of Environment, Forest, and Climate Change conducts awareness programs and policy initiatives to find effective solutions. However, the government alone cannot reach all 1.45 billion people in India. Therefore, social organizations and NGOs play a crucial role in this movement.

Numerous social organizations in India and around

the world are dedicated to mitigating climate change. By joining these organizations, you can contribute to environmental protection and support the planet's sustainability.

By joining organizations focused on mitigating climate change, you're contributing to a collective impact, driving systemic change, and helping to support a sustainable future.

142. Involve the Workforce for Climate Action to Mitigate Climate Risks and Drive Sustainable Solutions

Rising temperatures due to climate change are increasing health and economic risks for a significant portion of the global population, including India. Workers exposed to outdoor or hot indoor environments are at a heightened risk of heat stress and other heat-related disorders. Excessive heat can reduce work capacity and overall productivity.

Currently, national climate and employment policies often overlook the impact of climate change on the health and productivity of the workforce, with individuals adapting to environmental conditions as best they can.

Climate action is crucial to protect the workforce from the adverse impacts of climate change. Involving the workforce in environmental initiatives is a proactive step toward sustainability. You can encourage workforce participation in activities such as weekend plantation drives in your city and ensure the ongoing care of these trees as they grow. Additionally, guiding employees to reduce carbon emissions in daily life can empower them to contribute to a sustainable future.

By involving the workforce in climate change

mitigation efforts, you tap into a powerful resource, drive innovation, and create a sustainable future.

143. Speak to Drive Climate Action and Raise Awareness

Climate change impacts, such as extreme weather, pollution, and health issues, affect us all. If you're feeling the effects, why wait?

It's time to speak up! Share your concerns with friends, family, and elected officials. Raise awareness and contribute to solutions for a sustainable future.

Here are ways to use your voice:

- Share accurate climate information.
- Express concerns to leaders and policymakers.
- Support climate advocates and activists.
- Engage in conversations and social media.
- Write letters and participate in petitions.
- Attend public hearings and town hall meetings.
- Support climate-focused media and peaceful protests.
- Advocate for climate education.

By speaking up, you're driving action, inspiring change, and helping to create a sustainable future.

144. Raise Your Voice for Climate Action at Town Hall Meetings

Town halls are essential events where elected representatives, government officials, and social organizations gather to address various community issues, including health, education, agriculture, infrastructure, and food security. However, relatively few town halls are dedicated to environmental awareness, despite the increasingly severe impacts of climate change.

You can participate in town halls in your village or city and raise your voice for climate action. Share your involvement and the town hall details on social media to spread awareness. Your participation will inspire friends and community members to take individual actions for the environment. Additionally, it encourages elected representatives and government officials to prioritize and implement sustainable solutions.

Discuss topics such as clean energy, tree planting, clean fuels, reducing plastic pollution, and promoting organic farming with less reliance on chemical fertilizers.

By attending a town hall and raising your voice, you're driving climate action, promoting sustainability, and working toward a better future!

145. Use Your Voice: Write to Elected Representatives and Corporations for Climate Action

Elected representatives serve as custodians of the people in their constituencies. During elections, they engage with the community, listen to concerns, and commit to addressing issues once in power. Climate change, however, is a pressing concern that poses a significant threat to the survival of all living beings, including humans. It requires urgent attention and action.

As responsible citizens, writing letters to your elected representatives about environmental issues can make a significant impact. You can address concerns such as tree plantation in villages and city surroundings, clean energy, clean water, sanitation, proper waste management, reducing plastic use, and minimizing the use of chemical fertilizers. When representatives receive numerous letters from constituents, it encourages them to prioritize environmental concerns and explore solutions to mitigate the issues raised.

Writing letters to elected representatives and corporations can help mitigate climate change in the following ways:

Elected Representatives:

- Demand Climate Action: Encourage representatives to support climate policies and legislation.
- Share Climate Concerns: Express concerns about climate change impacts and its effects on your community.
- Support Renewable Energy: Advocate for transitioning to renewable energy sources.
- Encourage Climate Leadership: Urge representatives to take bold climate action.

Corporations:

- Demand Sustainable Practices: Push companies to adopt sustainable practices and reduce emissions.
- Support Climate-Friendly Policies: Encourage corporations to back climate-friendly policies.
- Invest in Renewable Energy: Advocate for companies to invest in renewable energy.
- Promote Sustainable Products: Urge companies to develop and promote sustainable products.

Tips for Effective Letters:

- Be clear and concise.
- Use personal experiences and stories to make the letter relatable.
- Focus on solutions and specific actions.
- Include specific demands or requests.
- Follow up with further communication if needed.

For Example:

Dear [Representative/CEO],

I'm writing to express my concern about climate change and its impacts on our community. I urge you to [specific demand or request]. I believe that [specific solution or action] can make a significant difference.

Thank you for considering my request.

Sincerely,

[Your Name]

By writing letters, you're using your voice to drive climate action, promote sustainability, and create a better future!

146. Leverage Legal Action: File Petitions in Court for Climate Change

With the rapid advancement of science and technology and a growing population, we have caused significant and often irreparable damage to our planet. Industries often overlook environmental standards due to inadequate monitoring. Climate change remains a neglected issue in the Indian Parliament, as representatives focus more on immediate concerns.

India is committed to various international climate agreements, including the Paris Agreement, but still faces challenges in achieving its goals, such as increasing forest cover to 33%. There is a pressing need for greater public awareness and action on climate change.

Filing petitions in court can create substantial pressure on the government and bring climate change to public attention. When multiple petitions are submitted, the government is more likely to address environmental protection schemes and initiate discussions in Parliament. For instance, a petition filed in the Supreme Court by a lawyer advocating for a climate emergency has prompted government responses. The more petitions that are filed, the greater the awareness and momentum for climate action.

Additionally, organizations like Change.org are collecting signatures for online petitions urging the government to declare climate change an emergency in India. The petitions are directed at the Ministry of Environment, Forests, and Climate Change, as well as the Prime Minister of India.

Filing petitions requires careful planning, research, and collaboration with legal experts and environmental organizations to increase the likelihood of success.

By filing a petition, you're leveraging the legal system to drive climate action, protect the environment, and promote a sustainable future!

147. Make Your Vote Count: Get Politically Active for Climate Action

Panchayat, Corporation, and State elections are held throughout India each year. Political parties and independent candidates present their election manifestos, promising actions to be taken once in power. While some parties have started incorporating climate action targets into their manifestos, public awareness and demand for these initiatives remain limited.

You can play a significant role by pressuring political parties and candidates to prioritize climate action in their manifestos. Encourage greater public awareness and demand for climate-focused points in election manifestos. Vote for parties or candidates who prioritize and support climate action.

Ways to Get Politically Active:

- Register to vote
- Research climate-forward candidates
- Volunteered for climate-focused campaigns
- Contact representatives and leaders

- Participate in climate protests and rallies
- Support climate-focused organizations
- Stay informed and educate others

Remember, voting is a crucial step in shaping a sustainable future!

By getting politically active and voting, you can use your voice to demand climate action, influence policy, and ensure a liveable planet for future generations!

148. Get Inspired: Watch an Environmental Documentary and Take Action for Climate Protection

Climate change, deforestation, species extinction, air pollution, plastic pollution, and water pollution are some of the most pressing issues of our time. These challenges affect every country and every living creature, including humans. Many storytellers worldwide are shedding light on these issues through powerful films.

Just as you watch movies and TV shows for entertainment and relaxation, environmental documentaries offer a unique way to deepen your understanding and emotional connection with nature. By watching these documentaries, you'll gain insight into the urgent need for climate protection and feel inspired to take action.

Countless environmental documentaries are available online; watch them to develop a closer connection to nature and understand how you can make a positive difference.

Get passionate, get informed, and take action to mitigate climate change!

149. Read Climate Fiction and Awareness Books: Fuel Your Passion for Climate Action

Many people are aware of climate change but feel helpless, believing there's little they can do about it. The daily demands of life often leave little time for reflection or action.

When you read novels—fiction or non-fiction—you choose topics that resonate with your interests and passions. Climate fiction and climate awareness books offer a way to connect with nature and inspire meaningful action. These books offer valuable knowledge and motivate and encourage you to take steps to protect the environment. They explore pressing questions such as:

- What's happening around us?
- Why is it happening?
- What drives extreme weather events?
- Why is the Earth's temperature rising?
- Why are cyclones becoming more destructive?
- How can we address rising pollution levels?

There are numerous climate fiction and awareness books available online. Choose the ones that speak to you, and they'll energize you to become a climate warrior. Additionally, your support will inspire more writers to focus on environmental topics.

Reading climate fiction and awareness books enhances climate literacy, promotes action, nurtures resilience, and fosters hope for a sustainable future.

Get reading, get informed, and take action to mitigate climate change!

150. Add Climate Change Books to Your Local Library: Foster Awareness and Inspire Action

In most Indian libraries, a wide range of fiction and nonfiction books are available, covering topics such as science, mathematics, history, geography, and novels on various subjects. However, due to limited awareness, books on climate change are rarely found.

You can contribute by purchasing a few climate change fiction and awareness books and suggesting to the librarian that they be displayed prominently in front of the library for easy access by all readers. Take the time to explain the ongoing climate crisis and the importance of collective action and solutions. Over time, more readers will engage with these books, fostering awareness about climate change and encouraging informed actions.

Your efforts will contribute to a more climate-conscious community. By supporting libraries with climate change literature, you empower people with essential knowledge, promote climate education, and inspire collective action for a sustainable future.

Take action, donate books, and help create a climate-informed community!

151. Support Climate Change Journalism: Promote Awareness and Foster Action

Many Indian and global publications focus on widely accepted topics, with limited attention to climate change books and journals. As a result, publishers receive fewer responses to climate-related content.

You can help by supporting climate-focused publications by purchasing books and journals that focus on climate change. An increase in readers will encourage publishers to prioritize climate change topics. Greater publication of climate-related materials will lead to increased awareness in society.

Ways to Support:

- Subscribe to climate-focused publications (e.g., Grist, Climate Progress).
- Donate to independent media outlets (e.g., Inside Climate News).

- Share climate articles on social media.
- Support investigative journalism organizations (e.g., ProPublica).
- Encourage mainstream media to prioritize climate coverage.

By supporting publications that report on climate change, you're fostering informed decision-making, promoting accountability, amplifying critical voices, encouraging climate action, and supporting vital journalism.

Take action, support climate journalism, and help create a more sustainable future!

152. Support Environmental Journalists: Amplify Climate Awareness and Inspire Action

Journalists play a vital role in a democratic society, keeping us informed about events. While mainstream journalists focus on sensitive, immediate issues, environmental journalists concentrate on climate crises, including air and water pollution, plastic pollution, deforestation, species extinction, rising sea levels, cleanliness drives, sanitation issues, and water scarcity. However, readers and viewers are often less engaged with climate-related news as these topics may not seem urgent in the short term. As a result, environmental journalists face challenges in gaining attention compared to other journalists covering immediate priorities.

We can support and encourage environmental journalists by engaging with their content. Reading their articles, watching their stories, liking, and commenting on their work helps motivate them to produce more impactful reporting. Increased reporting on climate issues leads to greater public awareness and action.

Ways to Encourage Environmental Journalists:

- Subscribe to their publications or podcasts.
- Share their articles and social media posts.
- Provide feedback and encouragement.
- Attend their events and webinars.
- Support organizations that fund environmental journalism.
- Collaborate with journalists on climate projects.
- Offer expertise or resources for their reporting.
- Celebrate their achievements and awards.

By encouraging environmental journalists, you foster critical storytelling, promote climate awareness and education, support accountability and transparency, inspire climate action and solutions, and strengthen the role of journalism in addressing climate change.

Take action, support environmental journalists, and help create a more sustainable future!

153. Encourage Your Friends to Act Sustainably: Creating a Ripple Effect for a Greener Future

One of India's ironies of climate change is that despite experiencing extreme heatwaves, rising temperatures, unusual cyclones, floods, and high pollution levels, people remain largely unaware of its impacts. You can engage with your friends about the ongoing environmental crises and how planting trees, managing waste through composting at home, reducing electricity usage, minimizing plastic consumption, growing vegetables, carrying reusable shopping bags, and choosing eco-friendly products can make a difference.

By discussing these topics with your friends and encouraging them to take sustainable actions, they will, in turn, influence others, creating a chain of awareness and change in society. This ripple effect supports the environment and promotes sustainable living.

Your small steps can lead to a significant societal impact. Together, let's work towards reducing our carbon footprint through simple yet effective actions, such as using public transportation, reducing meat consumption, and conserving energy. Every small action counts, and together, we can drive meaningful change!

154. Start a Climate Conversation: Inspiring Action and Building Awareness for a Sustainable Future

Humanity's greatest challenges are climate change, our mindset, and the actions we take in our daily lives. No reduction in carbon footprint will occur unless we discuss the issue personally, professionally, and within various forums. Everyone, including you, your family, friends, colleagues, business partners, neighbors, elected representatives (MLA, MP, Ministers, CM, PM), bureaucrats, judges, and even the President, is affected by the impacts of climate change. So, why wait? Start discussing the climate with everyone whenever you have the chance.

Initially, people may not respond, but they will eventually listen and respect your perspective.

Starting a climate conversation can help mitigate climate change by:

- Raising awareness: Educating others about the impacts of climate change and its solutions.
- Building empathy: Encouraging understanding and connection with those affected.
- Inspiring action: Motivating individuals to adopt sustainable habits and advocate for change.
- Creating a sense of community: Fostering a shared sense of purpose and responsibility.
- Encouraging collective action: Mobilizing individuals to work together towards a common goal.

Conversation starters:

- What do you think about the recent climate report?
- Have you noticed any changes in the weather lately?
- How do you think we can reduce our carbon footprint?
- What's your favorite way to live sustainably?
- How can we support each other in making eco-friendly choices?

Tips for effective conversations:

- Listen actively and empathetically.
- Avoid blame or judgment.
- Focus on solutions and actions.
- Share personal experiences and perspectives.
- Encourage open-mindedness and curiosity.
- Provide credible information and resources.
- Foster a sense of hope and empowerment.

Remember, every conversation counts; collective action can lead to significant positive change!

Start a climate conversation today and inspire others to join the movement towards a more sustainable future!

155. Be a Climate Volunteer: Contributing Your Time to Protect Our Planet

We are responsible for the damage to our planet and have the power to protect it. The world is filled with natural resources and beauty that we must safeguard rather than exploit. The climate continues to deteriorate due to our daily actions, and we are responsible for preserving it.

If you have some free time, consider dedicating it to climate volunteering. Climate volunteering is an excellent way to connect with new people and discuss ongoing climate crises, especially if you are new to an area. It strengthens your ties to the community and expands your support network by exposing you to individuals with shared interests, local resources, and engaging, fulfilling activities. Given that nature holds a deep significance for humanity, these experiences can be truly rewarding.

Numerous organizations are working tirelessly to protect the environment. You can become a climate volunteer based on your area of interest—whether it's plantation management, wildlife protection, waste management, renewable energy, coral protection, combating environmental pollution, promoting sustainable agriculture, forest conservation, supporting the Clean India mission, or river clean-ups, among others.

Take action, volunteer, and be part of the climate solution!

Every hour counts, and collective efforts can lead to significant positive change!

156. Become a Climate Warrior: Take Action and Lead the Fight Against Climate Change

The effects of climate change are devastating social and economic developments in India and around the world. Sustainable climate actions present opportunities to unlock the country's significant social and economic growth.

As a Climate Warrior, you'll gain a deeper understanding of the ongoing climate crisis and how empowering individuals can help protect the planet. By taking action, you'll create an identity recognized at the state, national, and even international levels as climate awareness grows. People will recognize you for your efforts to combat climate crises, and the sense of fulfilment from protecting nature will bring immense happiness. Humans are inherently connected to nature, and the beauty of life lies in preserving it.

You can find documentaries about climate warriors

online that inspire and motivate you. Be a climate warrior and lead the way to a sustainable future.

UNICEF, a renowned social organization, runs a Climate Warrior program for children in India.

Here are some ways to become a Climate Warrior and help mitigate climate change:

- Educate yourself: Learn about the causes, impacts, and solutions of climate change.
- Reduce your carbon footprint: Make eco-friendly lifestyle choices.
- Raise awareness: Share climate change information with others.
- Advocate for change: Support climate policies and initiatives.
- Participate in activism: Join climate protests, campaigns, and movements.
- Support renewable energy: Invest in solar, wind, and other clean energy sources.
- Conserve resources: Reduce, reuse, and recycle to minimize waste.
- Plant trees and support reforestation: Help absorb carbon dioxide.
- Support climate-friendly policies: Contact representatives and vote for climate action.
- Inspire others: Encourage friends, family, and community to take action.

Remember, every small action counts; collective efforts can lead to significant positive change!

Become a Climate Warrior and join the fight against climate change!

157. Donate for Climate Protection: Support Initiatives That Drive Positive Change

Supporting a cause you care about through donations benefits the recipient and can be deeply rewarding for you. Millions of people regularly contribute to charities they believe in, motivated by the positive impact on the cause and their well-being.

Today, the devastating effects of climate change necessitate your support in finding effective solutions. While governments are making efforts, public participation is crucial for achieving the desired results as individual actions continue to harm the environment. Numerous social organizations and NGOs work tirelessly to protect the climate through various initiatives. These organizations rely on public donations to sustain their efforts, making your contribution essential.

You can support the environment by donating some of your income or savings to organizations dedicated to climate protection. By selecting an area of interest and researching social organizations online, you can ensure that your contribution supports initiatives that align with your values. Your support will directly contribute to preserving the environment and driving meaningful change.

Every donation counts, and collective efforts can lead to significant positive change! Donate today and support the fight against climate change!

158. Choose Online Deliveries Judiciously: Reducing Waste and Lowering Carbon Footprint

With the rapid growth of online platforms such as Amazon India, Flipkart, Snapdeal, Alibaba, Myntra, IndiaMart, Nykaa, and FirstCry, many prefer to purchase items online by selecting from various options. Often, individuals browse these platforms for the best deals, purchasing items that may not be essential. This leads to over-ordering and the delivery of unnecessary products. Additionally, small single-item orders often result in excessive packaging, including plastic and other materials, which can harm the environment.

The production of packaging materials consumes energy, and transporting these products involves burning fossil fuels, which emits greenhouse gases. While it's convenient to shop online, your thoughtful and mindful purchases can significantly reduce your environmental impact.

By making informed choices and minimizing unnecessary orders, you can help reduce the carbon footprint associated with online deliveries and contribute to a more sustainable future.

159. Advocate for Energy-Efficient Building Codes: Driving Sustainability and Reducing Carbon Footprint

Residential and commercial buildings are responsible for nearly 6% of India's electricity consumption. To promote energy conservation and harness building energy savings, the Indian government developed the Energy Conservation Building Code (ECBC) for commercial buildings.

Energy efficiency codes are crucial for ensuring the sustainable construction and operation of buildings. Potential reductions in energy use range from 30% to 40%. Advocating for the broader adoption and implementation of the ECBC is crucial for conserving energy, improving air quality, and advancing a low-carbon future that aligns with India's climate goals.

You can support the implementation of energy-efficient building codes nationwide and push for new constructions to adhere to these standards. Additionally, monitoring residential constructions for energy efficiency is essential for long-term sustainability.

Remember, every voice counts; collective efforts can lead to significant positive change!

160. Spread Climate Awareness: Empowering Change for a Sustainable Future

Climate awareness is essential for achieving sustainable development in India and worldwide. It empowers people to understand and address the impacts of global warming, enhances climate literacy among young people, fosters favourable changes in attitudes and behaviours, and supports the preparation and implementation of climate adaptation measures nationwide. A lack of awareness contributes to environmental damage and creates barriers to effective climate adaptation.

You can be key in spreading awareness about the causes, effects, and solutions to combat climate change within your community. Sharing information through social media, engaging in discussions, and involving friends and family effectively educates and inspires others.

Spreading awareness to mitigate climate change can be done through:

- Social Media: Share posts, videos, and articles.
- Public Talks: Organize or participate in events, lectures, and discussions.
- Education: Integrate climate change into school curricula and workshops.
- Community Engagement: Involve local communities in climate initiatives.
- Collaborations: Partner with influencers, organizations, and businesses.

- Visual Media: Use films, documentaries, and art to convey the message.
- Personal Conversations: Have discussions with friends, family, and colleagues.
- Online Campaigns: Launch or support petitions and initiatives.
- Events: Host or participate in climate-related events, rallies, and protests.
- Storytelling: Share personal experiences about climate impacts.

Remember, awareness is the first step towards action! Spread awareness and inspire others to take a stand against climate change!

161. Let Policymakers Know You Are Concerned About Climate Change

India has the potential to develop sustainably without harming its natural resources in pursuit of ambitious development goals. It is essential for elected officials and business leaders to understand that climate action is urgent and necessary to save our planet. By raising awareness through various forums, including social media, policymakers and elected representatives will recognize your concern for climate issues. Increased public awareness will push policymakers to prioritize and allocate more resources towards climate actions.

Here are ways to let policymakers know you're concerned about climate change:

- Contacting Representatives: Contact your elected officials via calls, emails, or letters.
- Signing Petitions: Join or support online or offline petitions demanding climate action.

- Participating in Town Halls: Attend public meetings and ask questions related to climate change.
- Writing Letters to the Editor: Express climate concerns in local newspapers.
- Engaging in Social Media: Use hashtags, tag officials, and share climate-related content.
- Joining Advocacy Groups: Support organizations advocating for climate policies.
- Attending Climate Rallies: Participate in peaceful protests and demonstrations.
- Voting: Elect representatives committed to sustainable climate action.
- Providing Public Comments: Offer feedback on policy proposals and public consultations.
- Meeting with Officials: Arrange meetings to discuss your climate concerns.

Remember, policymakers respond to constituent concerns. Make your voice heard and demand climate action from your representatives!

162. Have Hope to Mitigate Climate Change: Driving Action for a Sustainable Climate Future

Climate change awareness is growing rapidly in India and around the world. World leaders have shown concern, with 195 countries attending the United Nations Climate Change Conference in Paris in December 2015. The goal was to limit global temperature rise to well below 2°C above pre-industrial levels by reducing greenhouse gas emissions. In 2016, 189 countries signed the Paris Agreement.

However, in 2017, the U.S. withdrew from the agreement under President Donald Trump. Later, on January 20, 2021, President Joe Biden re-joined the Paris Agreement.

Climate change remains a critical issue. It is essential to take steps in our daily lives to reduce carbon emissions, while also hoping for more decisive climate action from governments worldwide.

Having hope is crucial because:

- Hope inspires action: Believing in a better future motivates us to take steps toward it.
- Hope fosters resilience: Staying positive helps us manage climate-related challenges.
- Hope encourages collective action: Shared optimism unites people to achieve common goals.
- Hope drives innovation: Believing in solutions fuels creativity and new ideas.
- Hope promotes sustainability: Focusing on a hopeful future encourages sustainable practices.
- Hope supports mental health: Climate anxiety can be overwhelming; hope helps maintain well-being.
- Hope enables climate justice: Believing in a fairer future ensures equitable resource distribution.
- Hope bridges generations: Working together for a shared future fosters intergenerational cooperation.
- Hope celebrates successes: Recognizing progress motivates continued climate action.
- Hope creates a better future: Believing in positive outcomes drives meaningful change.

Remember, hope is a powerful catalyst for change. Hold onto hope, take action, and work towards a sustainable future!

163. Recognizing Climate Champions: Celebrating Action for a Sustainable Future

Felicitation for Climate Champions is an excellent idea, but it is essential to recognize and appreciate individuals, organizations, and communities that make significant contributions to climate action and sustainability.

Felicitation for climate champions is essential to:

- Recognize their efforts: Acknowledge their dedication and hard work.
- Inspire others: Showcase their achievements to motivate others to take action.
- Encourage continued action: Celebrate their progress to keep them engaged.
- Build a sense of community: Unite climate champions and foster a shared sense of purpose.
- Amplify their impact: Increase their influence by highlighting their accomplishments.
- Provide a platform: Offer a stage for climate champions to share their stories and ideas.
- Foster collaboration: Connect climate champions to leverage their collective expertise.
- Drive accountability: Encourage climate champions to continue their work.
- Celebrate milestones: Mark essential achievements in the fight against climate change.
- Keep the momentum: Sustain the energy and enthusiasm for climate action.

Ways to felicitate climate champions include awards and honours, Public recognition, Social media spotlights, Feature stories and interviews, Community events and celebrations, Leadership opportunities, Networking platforms, Capacity-building programs, Collaborative projects, and Tokens of appreciation (e.g., certificates, plaques).

Remember, recognizing climate champions inspires a wave of positive change!

164. Advocate for Sustainable Public Transport Solutions: A Path to Cleaner, Greener Cities

The transport sector is a significant contributor to global greenhouse gas emissions, with road transport in India playing a significant role, according to the International Energy Agency (IEA). Private vehicles emit an average of 4.6 metric tons of CO_2 annually, accounting for a significant portion of the transport sector's 13% contribution to India's total emissions. Public transport, however, offers a powerful alternative, reducing carbon emissions by up to 90% per passenger compared to private vehicles. In addition to lowering emissions, public transit enhances energy efficiency, decreases air pollution, and fosters sustainable urban planning.

Promoting diverse modes of public transport—such as buses, trains, subways, light rail, cycling infrastructure, walking paths, and car- or ride-sharing services—can revolutionize mobility systems while ensuring equitable access for all. Advocacy efforts to improve public transportation can include engaging with local authorities, participating in public consultations, organizing community campaigns, collaborating with environmental groups, and utilizing social media to raise awareness.

Supporting better public transport solutions helps reduce emissions, optimize energy use, improve urban mobility, and combat climate change, creating a more sustainable future for everyone.

165. Switch to Climate-Conscious Banks: Financing a Sustainable Future

Access to finance is a critical factor in transitioning to a low-carbon economy. Green and sustainable lending ensures that capital is directed toward projects and initiatives that have a positive impact on the environment. Achieving global net-zero emissions by 2050 will require an estimated USD 200 trillion in investment, making the banking industry pivotal to this transformation. Banks, through their financing decisions, have the power to determine which businesses and projects receive capital, significantly influencing the fight against climate change.

Promising signs of progress include stricter lending criteria for high-carbon industries, increased funding for renewable energy and green technologies, and partnerships between banks, governments, and development institutions to promote climate finance. Additionally, many banks now offer green loans with favorable terms for environmentally beneficial projects to meet their sustainability goals. However, some banks remain slow to adopt climate finance initiatives, making it essential for individuals to act responsibly by switching to institutions that prioritize sustainability.

Switching to climate-conscious banks can significantly contribute to mitigating climate change. It redirects funding to climate-friendly projects, encourages sustainable banking practices, supports the development of renewable energy, reduces greenhouse gas emissions, and promotes climate-responsible investing.

Key Criteria for Choosing a Climate-Friendly Bank:
- Investment policies favouring sustainability
- Commitment to fossil fuel divestment
- Financing for renewable energy projects
- Support for sustainable agriculture
- Strong Environmental, Social, and Governance (ESG) principles

Steps to Make the Switch:
- Research banks with climate-friendly policies

- Move your accounts to institutions with robust ESG practices
- Support Community Development Financial Institutions (CDFIs)
- Invest in climate-focused funds
- Participate in campaigns holding banks accountable for their environmental impact

By switching to a bank prioritizing climate finance, you can help drive funding toward sustainable initiatives, foster greener banking practices, and contribute to a global effort to combat climate change.

166. Buy from companies that make efforts to reduce their emissions

India's commitment to the Paris Agreement has pressured businesses to adopt sustainable practices and reduce greenhouse gas emissions across production, distribution, and supply chains. While some companies have adopted these changes, many have yet to take action. As a consumer, you can play a pivotal role by supporting companies that prioritize emission reduction.

Buying from such companies encourages sustainable practices, promotes the adoption of green technologies, influences industry-wide change, and ultimately reduces the overall carbon footprint. Look for businesses that adopt renewable energy, improve energy efficiency, manage sustainable supply chains, offset carbon emissions, and develop eco-friendly products.

Simple actions include researching companies' sustainability efforts, opting for products with minimal packaging, supporting carbon-neutral certified brands, and advocating for transparency in emissions data.

By making mindful purchasing decisions, you can

drive positive change and help mitigate the effects of climate change.

167. Embracing Sustainable Agriculture for Climate Resilience and Food Security

A growing global population and changing diets have intensified the demand for food, with an estimated 70% more food needed by 2050 to feed 9 billion people. While essential, agriculture contributes 19–29% of global greenhouse gas emissions and faces significant challenges from climate change, including declining crop yields and increased economic losses from natural disasters.

In India, climate-sensitive agriculture is particularly vulnerable, with significant crop yields projected to decrease by 10–40% by 2100 due to temperature rise, rainfall variability, and limited irrigation.

Adopting sustainable agriculture practices is essential to address these challenges. Sustainable methods can sequester carbon, reduce emissions, improve soil health, enhance water management, and strengthen food security, mitigating the impact of climate change while ensuring agricultural resilience.

168. Start Climate Conversations with Family and Friends for a Greener Future

Discussing climate change with your family and friends is vital to protecting our planet and promoting awareness. Although these conversations sometimes feel challenging, employing thoughtful strategies can make them effective and meaningful. You can foster understanding and inspire action by asking open-ended questions, sharing facts and personal stories, and connecting climate issues to topics that matter to your loved ones.

Tips for Explaining Climate Change:

- Ask about their perspectives to gain a deeper understanding of their views.
- Use real-life examples and data to illustrate the impacts of climate change.
- Relate climate change to their interests and values.
- Share how climate change has affected your life.
- Suggest actionable solutions they can adopt.

Guidelines for Meaningful Conversations:

- Choose a relaxed moment to bring up the topic.
- Keep a positive tone and avoid overly complex jargon.
- Share personal experiences and the broader impacts of climate change.
- Actively listen and find common ground.
- Present solutions alongside the challenges.
- Advocate for impactful changes, both personal and systemic.
- Encourage eco-friendly habits and leave room for follow-up discussions.

By initiating these conversations, you empower your loved ones to educate themselves, take action, and amplify the environmental movement. Together, these discussions can strengthen our collective response to climate change challenges.

169. Take a Stand Against Deforestation for a Healthier Planet

Every year, 10 million hectares of forests—an area the size of Portugal—are cleared, devastating wildlife, ecosystems, and billions of people who depend on forests for food, water, and livelihoods. Deforestation also exacerbates climate change, with tropical tree loss releasing over 5.6

billion tonnes of greenhouse gases annually, exceeding the combined emissions of aviation and shipping.

Forests play a critical role in stabilizing the climate by absorbing carbon dioxide through photosynthesis and storing it. However, deforestation releases this stored carbon and reduces the planet's ability to absorb future emissions, compounding the climate crisis.

Addressing deforestation with climate change, driving extreme weather events and massive economic losses, is crucial.

By halting deforestation, India can work toward its target of 33% forest and tree cover (up from 24.62% in 2021) and fulfil its climate commitments under the Paris Agreement. Protecting forests is essential for preserving biodiversity, supporting livelihoods, and combating climate change.

170. Shift the Mindset from Individual to Collective Action for Greater Climate Impact

Addressing climate change requires moving beyond individual actions to embrace collective efforts. While individual actions, like walking instead of driving, reflect personal decisions, collective actions involve groups working together toward shared goals, such as a neighbourhood installing sidewalks to promote walking. Collective actions are more impactful due to their larger scale and shared resources.

Shifting the mindset from individual to collective action is essential for reducing greenhouse gas emissions and combating climate change. Collective efforts amplify impact through shared initiatives, foster community-led solutions, drive policy changes, promote cultural shifts toward sustainability, and strengthen global cooperation.

Effective strategies for collective action include community-based projects like renewable energy cooperatives, advocacy campaigns like climate strikes, and policy initiatives like carbon pricing, educational programs for climate literacy, and collaborative research and development. Simple contributing steps include joining local climate groups, participating in community initiatives, supporting climate policy advocacy, engaging in climate education, and collaborating on climate-related projects.

By embracing collective action, we can amplify efforts, encourage sustainable community practices, drive systemic change, and effectively mitigate climate change.

171. Advocate for Stronger Climate Action from Elected Officials

Pressuring elected officials is a crucial step in combating climate change. By urging leaders to prioritize climate policies, we can drive reductions in carbon emissions, increase investment in renewable energy, and promote the adoption of sustainable practices.

Why Public Pressure Matters:

When citizens voice their concerns, governments are more likely to take action on climate change, providing benefits such as enhanced energy security, improved public health, and reduced pollution.

How to Make a Difference:

- Contact Representatives: Call, email, or write to express your concerns about climate issues.
- Join Advocacy Groups: Collaborate with organizations pushing for climate action.
- Attend Climate Events: Attend rallies, protests, and town halls.

- Vote for Climate-Conscious Leaders: Support candidates committed to addressing climate challenges.

What We Can Achieve Together:

By uniting in collective action, we can reduce emissions, advance renewable energy, and support climate-resilient infrastructure and education. Every voice matters—demand climate action from our leaders today!

172. Contribute to a Net-Zero Workplace for a Sustainable Future

Climate change is creating widespread disruptions globally, including rising temperatures, prolonged droughts, intense heatwaves, and extreme weather events. According to the Intergovernmental Panel on Climate Change (IPCC), these challenges will escalate unless greenhouse gas emissions are drastically reduced.

A net-zero workplace is crucial in addressing this crisis by integrating sustainable practices and low-emission technologies. Employees are key contributors to this effort, adopting energy-efficient solutions, incorporating greenery into workspaces, and helping organizations achieve their net-zero goals.

A net-zero workplace offers numerous benefits, including reduced greenhouse gas emissions, enhanced energy efficiency, improved brand reputation, support for sustainable development, and increased employee engagement. Strategies to achieve this include adopting renewable energy sources such as solar and wind, utilizing energy-efficient lighting and equipment, retrofitting buildings with green design features, promoting sustainable transportation options like electric vehicle charging, reducing waste, tracking emissions, and educating employees on sustainability.

Simple actions, such as conducting energy audits, reducing paper use, implementing energy-efficient practices, promoting sustainable commuting, and involving employees in sustainability initiatives, can have a significant impact.

Contributing to a net-zero workplace helps reduce emissions, improve energy use, and support global efforts to combat climate change.

173. Boycott Major Polluters for a Cleaner Future

Consumer choices have the power to significantly impact climate change mitigation. By prioritizing sustainability, individuals can make informed decisions about the companies they support when purchasing groceries, clothing, household items, building materials, electronics, and more. It's crucial to differentiate between companies that contribute heavily to pollution and those that adopt sustainable production, distribution, and supply chain management practices.

Choosing not to buy from major polluters is a form of accountability. It pressures these companies to adopt sustainable, low-emission practices in order to achieve their net-zero goals. This shift reduces the demand for polluting products, encourages cleaner energy transitions, decreases greenhouse gas emissions, and promotes corporate responsibility and environmental stewardship.

174. Embrace Renting Over Buying for a Greener Future

Renting offers a practical and eco-friendly alternative to purchasing for occasional needs, such as dresses for events or cars for infrequent trips. Renting allows multiple people to share resources, reducing waste and the need for storage while being cost-effective. Beyond clothing and cars, various items such as furniture, tools, and electronics are also available for rent, offering a sustainable solution for different needs.

Adopting a "rent instead of buy" mindset reduces consumption, minimizes waste, lowers resource extraction and emissions, encourages collaboration, and supports a circular economy.

You actively contribute to environmental protection and climate change mitigation by choosing to rent.

175. Educate Your Children on the Value of Nature and the Environment

Raising environmentally conscious children is key to building a sustainable future. Teaching them about the importance of nature helps inspire lifelong values of environmental stewardship. Engage them with fun activities like nature scavenger hunts, art projects using natural materials, gardening, and camping to foster a love for the outdoors. Introduce them to animals, show them sustainable habits like conserving water and reducing waste, and involve them in community efforts, such as picking up trash at parks.

Encourage outdoor exploration, teach them about climate change, and use books or videos to inspire their curiosity about the planet. By fostering eco-friendly habits and raising awareness, we empower children to protect the environment, contribute to sustainable communities, and combat climate change.

176. Advocate for Clean Electricity with Your Utility Provider

Supporting renewable energy through your utility provider can significantly impact the transition to a greener future. In India, consumers can purchase renewable energy (RE) through distribution companies or private suppliers under the Green Energy Open Access Rules (GEOAR) introduced in 2022. This initiative enables consumers with a contract demand of 100 kW or higher to opt for renewable energy (RE) at a green tariff.

Private companies, such as Tata Power and Adani Electricity, in Mumbai already offer green power options. Tata Power reports that green electricity reduces 200,000 tonnes of carbon emissions annually. While adoption is still early, consumer demand can increase availability and lower costs.

By choosing clean electricity, you help reduce emissions, support renewable energy production, and mitigate climate change, making a meaningful contribution to a sustainable future.

177. Drive Sustainability Initiatives at Your Office

Incorporating sustainability into the workplace is essential for addressing climate change and fostering long-term success. By adopting eco-friendly practices into their office culture, businesses can reduce their carbon footprint, enhance brand reputation, and support sustainable development.

Key Steps to Promote Sustainability in the Office:

- Cultivate a Green Culture: Motivate employees with initiatives like plastic-free lunches, bike-to-work days, reusable item campaigns, or green challenges with rewards for participation.
- Adopt a Paperless Approach: Transition to digital documentation, use recycled paper, and set printers to double-sided and toner-saving modes.
- Encourage Eco-Friendly Commuting: To reduce

commuting emissions, promote carpooling, public transportation, biking, or remote work options.

- Create a Healthy Workspace: Incorporate plants to improve air quality, boost morale, and explore sustainable office design options.
- Conserve Energy: Turn off unused electronics, optimize heating and cooling systems, and install energy-efficient lighting.

Supporting corporate sustainability benefits the environment, improves operational efficiency, and builds a responsible brand image. Start small and involve everyone in making meaningful changes. Together, we can create a lasting impact.

178. Donate Used Books to Public Libraries for a Sustainable Future

Donating used books to public libraries is a simple yet impactful way to promote literacy, build community connections, and support environmental sustainability. It provides access to diverse knowledge, fosters education, and reduces waste by extending the life of books.

Your contributions can inspire learning, comfort those in need, and help cultivate a culture of sharing and responsible consumption. By giving your books a second life, you declutter your space and make a meaningful difference in others' lives while supporting sustainable practices.

Donating used books can reduce waste and landfill contributions, conserve resources, promote sustainable literacy, support community sharing, and mitigate climate change.

179. Clean Up Your Digital Footprint for a Greener Future

Every time you use the internet—whether it's social

media, personal or company emails, Google searches, or watching videos and newsletters—you contribute to CO2 emissions. While stopping the use of the internet is unnecessary, you can make mindful choices to reduce unnecessary use. Unsubscribe from irrelevant newsletters, delete unimportant emails, and move spam to the appropriate folder.

Digital footprint reduction strategies include deleting unused emails and files, unsubscribing from unwanted newsletters, utilizing cloud storage efficiently, disabling unnecessary device features, and selecting energy-efficient devices and hardware.

By managing your digital footprint, you can reduce energy consumption and e-waste, decrease carbon emissions, enhance cybersecurity, promote sustainable digital practices, and help mitigate climate change.

180. Drive Smart, Drive Green: Avoid High-Speed Driving

Every vehicle has an optimal speed range that minimizes fuel consumption, which varies depending on the car's type, design, and age. Fuel consumption significantly increases beyond 80 km/h due to increased air resistance. However, speed is not the only factor. Regardless of your car, you can reduce fuel consumption—and therefore emissions—by driving smoothly.

This includes anticipating corners, avoiding sudden braking, easing off the accelerator before reaching the peak of a hill, and removing unnecessary roof racks or heavy items inside the vehicle to make it lighter and more streamlined.

Driving at 60-80 km/h instead of 120 km/h reduces emissions by 20-30%, while aggressive driving, such as speeding and rapid acceleration, increases emissions by 30-40%. Fuel-efficient driving practices alone save 2.4 billion litters of fuel annually.

By avoiding high-speed driving, you can reduce fuel consumption and emissions, enhance fuel efficiency, promote sustainable transportation habits, support climate-resilient infrastructure, and help mitigate the effects of climate change.

181. Stay Informed: Subscribe to Climate Web Resources

Climate knowledge is essential, especially in India, where awareness of Climate Change and its causes, impacts, and mitigation is still growing slowly. By subscribing to climate-related web resources, you can stay up-to-date with the latest news and information, helping you make informed decisions about potential mitigation strategies. Web Resources include:

- News outlets (e.g., Climate Change News, The Climate Report)
- Scientific journals (e.g., Nature Climate Change, Environmental Research Letters)
- Government reports (e.g., IPCC, NASA Climate Change)
- Non-profit organizations (e.g., World Wildlife Fund, The Climate Group)
- Online courses and educational platforms (e.g., Coursera, edX)

Simple actions include subscribing to climate-focused newsletters, following climate experts and organizations on social media, engaging with online communities and forums, participating in webinars and events, and sharing climate information.

By subscribing to web resources, you can stay

informed on climate issues and solutions, enhance climate literacy and awareness, support sustainable decision-making, encourage climate action and advocacy, and help mitigate climate change.

182. Support the Swachh Bharat Mission

The Hon'ble Prime Minister launched the Swachh Bharat Mission (SBM) on October 2, 2014. Its focus is on making India Open Defecation-Free (ODF). Over the past five years, this movement has led to the construction of over 100 million individual household toilets, with the goal of achieving 100% sanitation coverage by 2019.

By October 2019, more than 6 lakh villages had been declared Open Defecation Free (ODF), a significant milestone marking Mahatma Gandhi's 150th birth anniversary. Following this achievement, SBM-G Phase II was launched to sustain the ODF status and manage solid and liquid waste by 2024-25, transforming villages into ODF Plus Model villages.

SBM-G Phase II focuses on:

- Sustaining ODF status
- Implementing solid and liquid waste management
- Improving visual cleanliness

The total estimated outlay for SBM (G) Phase-II is Rs.1.40 lakh crores, with convergence between various Government schemes.

ODF Plus progress is categorized on SBM (G) MIS into:

- Aspiring: Sustaining ODF with arrangements for SWM or LWM
- Rising: Sustaining ODF with both SWM and LWM
- Model: Sustaining ODF with SWM, LWM, and visual cleanliness

Supporting Swachh Bharat Mission enhances sanitation, waste management, and sustainability, contributing to environmental protection and climate change mitigation.

183. Protect Wildlife and Their Habitats

Climate change is a significant threat to our planet, raising sea levels, causing extreme weather events, and causing biodiversity loss. Wildlife is particularly vulnerable, as changes in temperature, rainfall patterns, and habitats disrupt ecosystems and threaten species.

The loss of habitats due to climate change leads to habitat fragmentation, making it difficult for species to find food, shelter, and mates. This can result in population declines, an increased risk of extinction, and a reduction in ecological services that benefit humans.

Protecting wildlife and their habitats is crucial for preserving biodiversity, enhancing ecosystem resilience, and mitigating the effects of climate change.

184. Protect Marine Habitats and Ecosystems

The ocean is vital to life on Earth, providing oxygen and nutrients and regulating the planet's climate. However, human activities pose significant threats to marine life through plastic pollution, overfishing, habitat loss, ocean acidification, and global warming.

Single-use plastics harm marine habitats, while overfishing disrupts ecosystems. Habitat loss in seagrass meadows, mangroves, and coral reefs further depletes biodiversity. Ocean acidification, caused by excess carbon dioxide, impacts marine life.

Reducing plastic use, supporting eco-friendly seafood practices, participating in beach cleanups, and advocating for marine conservation policies are crucial for protecting marine ecosystems and mitigating the effects of climate change.

185. Protect Birds and Their Ecosystems

Birds are beautiful and fascinating creatures. They are vital for the health and balance of our ecosystems and provide numerous benefits to humans and other living beings.

Protecting birds is essential for maintaining healthy ecosystems and mitigating climate change:

- Pollination: Birds like hummingbirds and honeyeaters help pollinate plants, ensuring the reproduction of numerous species and maintaining biodiversity.
- Pest Control: Birds naturally control pests, reducing the need for harmful pesticides and supporting sustainable agriculture.
- Seed Dispersal: Birds help spread seeds across regions, aiding ecosystem restoration and promoting plant diversity.
- Carcass Clean-up: Birds such as vultures and crows clean up dead animals, preventing disease spread and recycling nutrients into the environment.
- Soil Fertilization: Bird droppings contribute to nutrient-rich soils, enhancing agriculture and supporting healthy ecosystems.

By protecting birds and their habitats, we promote ecosystem resilience, carbon sequestration, and climate regulation, helping to combat climate change.

186. Keep Our Beaches Clean: A Step Toward Coastal Conservation

Keeping our beaches clean is essential for preserving coastal and marine ecosystems. Clean beaches play a vital role in protecting marine life, maintaining natural habitats, improving water quality, and mitigating climate change impacts. They also support biodiversity, provide resilience against natural disasters, and offer safe nesting grounds for endangered species.

- Protection of Marine Life: Clean beaches reduce plastic and debris, preventing marine animals from ingesting or becoming entangled, which helps maintain healthy populations.
- Preservation of Natural Habitats: Clean beaches promote dune stability, support biodiversity, and reduce erosion, creating resilient coastal ecosystems.
- Water Quality Improvement: By minimizing pollutants, clean beaches ensure healthier marine ecosystems and reduce toxins in the marine food web.
- Mitigation of Climate Change Impacts: Healthy beaches, like mangroves and seagrass beds, absorb carbon dioxide and provide protection against natural disasters.
- Conservation of Endangered Species: Clean beaches offer safe nesting grounds for endangered species like sea turtles, helping to preserve biodiversity.

Cleaning beaches is crucial for enhancing ecological health and mitigating climate change. Collective efforts in beach conservation ensure a sustainable future for our coastal environments. Individual actions like participating in beach clean-ups, reducing plastic use, and supporting conservation organizations make a significant difference.

187. Implementing Energy-Efficient Heating and Cooling Systems for Sustainability

Using energy-efficient heating and cooling systems

reduces energy consumption, lowers utility bills, and minimizes environmental impact. Upgrading to high-efficiency HVAC systems with ENERGY STAR® certification ensures optimal performance. Proper system sizing, done by a professional load calculation, prevents energy waste. Heat pumps, including air-source and geothermal systems, are efficient choices as they transfer heat rather than generate it. Pairing these systems with a smart thermostat further boosts efficiency by learning your habits and optimizing energy use.

Regular maintenance, such as cleaning filters, sealing ducts, and annual inspections, ensures system efficiency. Enhancing insulation and air sealing helps maintain consistent indoor temperatures, reducing the HVAC workload. Zoning systems allow you to heat or cool only the used areas. For more significant savings, consider solar panels or solar water heating. Simple strategies like ceiling fans or natural ventilation can reduce reliance on mechanical systems. Take advantage of incentives and monitor your energy use to identify further improvements.

These steps create a sustainable, cost-effective heating and cooling solution.

188. Promoting Green Insurance Solutions for a Sustainable Future

Encouraging green insurance solutions promotes sustainability and rewards eco-friendly practices. Green insurance policies offer incentives for energy efficiency, renewable energy adoption, and sustainable living. Start by raising awareness of the benefits, such as discounts for energy-efficient appliances, solar panels, or electric vehicles, to encourage adoption. Highlight how these policies reduce premiums and contribute to a healthier planet.

Insurance companies can help by offering lower premiums for energy-efficient homes, coverage for renewable energy systems, or water conservation and waste reduction incentives. Collaborating with governments and environmental organizations can make green insurance more accessible, and governments can further promote it through tax incentives or subsidies.

Consumers can support green insurance by choosing insurers that prioritize sustainability and drive demand for eco-conscious options. These efforts can mainstream green insurance, helping build a more sustainable future.

189. Supporting Green Urban Infrastructure for Sustainable Cities

Advocating for green urban infrastructure is a powerful way to mitigate climate change and create more sustainable cities. Individuals can support initiatives like green roofs, urban forests, community gardens, and permeable pavements, which help reduce the urban heat island effect, improve air quality, and absorb carbon dioxide. To make an impact, people can participate in local government meetings, join sustainability-focused organizations, and support policies prioritizing green spaces and energy-efficient buildings.

Additionally, adopting eco-friendly practices such as planting trees, maintaining home gardens, using public transportation, and reducing energy consumption can contribute to greener urban environments.

Individuals can inspire collective action by raising awareness and actively engaging in community efforts, ultimately creating more resilient, sustainable cities for future generations.

190. Minimizing Chemical Waste in Agriculture

Reducing chemical waste in agriculture is essential for environmental protection, food safety, and sustainable farming. Overusing chemical fertilizers, pesticides, and herbicides can degrade soil, contaminate water, and harm ecosystems. Farmers can adopt integrated pest management (IPM) techniques, focusing on natural pest control, crop rotation, and resistant crop varieties. Precision agriculture technologies like GPS-guided equipment and soil sensors can optimize fertilizers and pesticides, applying them only when necessary. Transitioning to organic farming, which uses natural inputs such as compost and biological pest control, can further reduce chemical waste.

Governments and organizations can support these practices through education, training, and financial incentives for sustainable farming. Consumers can contribute by choosing organic and sustainably produced foods, driving demand for environmentally friendly agriculture.

By reducing chemical waste, we protect natural resources, promote biodiversity, and strengthen the resilience of our food system.

191. Promoting Sustainable Urban Heating Systems for a Greener Future

Advocating for sustainable urban heating systems is a crucial step in mitigating climate change and reducing the carbon footprint of cities. Traditional heating systems, which often rely on fossil fuels like coal, oil, and natural gas, are major contributors to greenhouse gas emissions. Transitioning to sustainable alternatives, such as district heating systems powered by renewable energy, geothermal heat pumps, or solar thermal systems, can significantly lower emissions and improve energy efficiency. Individuals can advocate for these changes by raising awareness about the benefits of sustainable heating systems, engaging with local policymakers, and supporting initiatives that promote clean energy infrastructure. Community-led projects, such as cooperative energy schemes or neighbourhood retrofitting programs, can also drive the adoption of greener heating solutions.

Additionally, individuals can make a difference by upgrading their homes with energy-efficient heating technologies, improving insulation, and using smart thermostats to reduce energy consumption. Governments and businesses play a key role by investing in research, offering subsidies for renewable heating systems, and setting regulations that phase out fossil fuel-based heating.

By advocating for and adopting sustainable urban heating systems, we can reduce reliance on fossil fuels, enhance energy security, and create healthier, more climate-resilient cities for future generations.

Business Level
"The environment should be our top priority because no business will survive without it."

> **Ratan Tata**, former Chairman of Tata Group

India's Journey towards Net Zero by Businesses
In November 2021, India announced its target to achieve net-zero emissions by 2070 at the 26th session of the United Nations Framework Convention on Climate Change (COP 26). As one of the fastest-growing economies, India's commitment is critical for meeting international climate goals. The corporate sector, responsible for approximately half of India's energy consumption, plays a vital role in this transition towards decarbonisation.

Indian corporates have begun voluntary actions to reduce emissions, driven by competitive pressure and investor demands. However, challenges like the absence of globally recognized standards, unclear policy frameworks, and nascent technologies make the path to net zero uncertain.

As of June 2022, 89 corporates have committed to science-based emission targets, participating in global initiatives like RE100, REDE, EV100, and CDP. Despite these efforts, only a small percentage of the world's largest companies are on track to reach net zero by 2050.

Surveys conducted by NSE 100 companies reveal:

Most companies see climate commitments as essential for competitive advantage and corporate reputation.

Renewable power and energy efficiency are the most preferred decarbonization strategies.

Barriers include technical and financial constraints, poor policy visibility, and lack of regulatory support.

A conducive policy framework and financial incentives are key enablers for climate action. PwC's report highlights that about 50% of Indian businesses have committed to achieving net-zero targets, with approximately 48% aiming for 2030. These commitments reflect a strong desire to transition to a low-carbon economy, although challenges persist in ensuring policy stability and making early financial investments.

India's corporate sector is making strides towards sustainability, but continued policy support and financing are essential to accelerate the journey towards net zero.

Let's discuss different mitigation points to be adopted by corporates for a NetZero Planet by 2050.

192. Net Zero Business

Developing and implementing policies for net-zero-compliant businesses is essential to addressing the climate crisis. By setting clear, actionable goals, businesses can significantly reduce their carbon footprint and contribute to global efforts to combat climate change.

Key Steps to Frame and Implement Net-Zero Policies:

- **Develop a Comprehensive Roadmap:** Establish a clear methodology for measuring and monitoring emissions aligned with international standards and create a phased approach with interim milestones, reflecting technological and economic viability.
- **Innovation and Pilot Projects:** Invest in emerging technologies and sustainable business models through pilot projects to explore low-carbon solutions.
- **Impact Assessment:** Assess the social, environmental, and financial impacts of transitioning across supply chains and business operations.
- **Training and Workforce Development:** Implement reskilling programs to build a workforce capable of supporting net-zero goals.
- **Collaboration and Knowledge Sharing:** Work with industry bodies, think tanks, and advisory groups like SBTi to optimize implementation and share best practices.
- **Monitoring and Reporting:** Regularly track progress, adapt strategies as needed, and report transparently to stakeholders.

Benefits of Net-Zero Policies:

- **Reduced Emissions**: Mitigating climate change by reducing corporate carbon footprints.
- **Enhanced Reputation**: Strengthening corporate image through sustainable practices.

- **Compliance and Regulatory Support**: Ensuring alignment with evolving climate regulations.
- **Innovation and Competitive Edge**: Leading sustainability initiatives and gaining a competitive advantage.
- **Stakeholder Trust**: Building trust with investors, customers, and the community.

By embedding net-zero policies into corporate operations, businesses can play a pivotal role in addressing climate change and fostering a sustainable, resilient future.

193. Corporate Environmental Responsibility Policy: A Strategic Approach to Environmental Protection and Climate Resilience

Corporate Environmental Responsibility (CER) is crucial for ensuring the environment's well-being, which supports every company's survival. With increasing awareness of environmental and societal impacts, CER has become an integral part of business strategy.

Importance of CER:

- **Sources of Greenhouse Gases**: Energy plants, residential buildings, industrial housing, agriculture, road transportation, and commercial buildings are top contributors.
- **Guidelines by Ministry of Environment**: Companies seeking green clearance must set aside up to 2% of capital investment for CER, brownfield projects require 0.125% to 1% of additional capital investment, and Greenfield projects allocate 0.25% to 2%.
- **CER Activities**: Infrastructure, such as drinking water, sanitation, health, roads, solar power, and waste management, is monitored through compliance

reports submitted to Regional offices and district collectors.

Fund allocation for Corporate Environmental Responsibility:

Capital Investment / Additional Capital Investment (In Rs.)	Green Field Project - % of Capital Investment	Brownfield Project - % of Additional Capital Investment
< 100 Cr	2.00%	1.00%
> 100 Cr to < 500 Cr	1.50%	0.75%
> 500 Cr to < 1000 Cr	1.00%	0.50%
> 1000 Cr to < 10000 Cr	0.50%	0.25%
> 10000 Cr	0.25%	0.125%

Benefits of CER:

- **Environmental Protection**: Reducing negative impacts on nature.
- **Climate Resilience**: Helps mitigate climate change through sustainable practices.
- **Business Success**: Enhances reputation and stakeholder trust.

This is a very good policy for protecting the environment for businesses that affect the environment during project construction. Effective implementation of the CER policy will significantly boost the environment and is one of the resilience measures to Climate Change.

A firm CER policy drives business success while safeguarding the environment!

194. Internal Carbon Pricing: A Strategy for Climate Action and Sustainable Growth

Internal carbon pricing is a key corporate climate action that fosters innovation, sustainability, and long-term competitiveness. It helps companies identify climate risks and inefficiencies, preparing them for a low-carbon economy.

By integrating carbon costs into investment decisions, companies improve their return on investment and align with global climate goals. Science-based targets (SBT) complement this approach by setting emission reduction targets that limit global warming to below

Concepts like internal carbon pricing and science-based targets are new and challenging to implement. Only a regulatory push can pressure companies to do more for the climate.

Internal carbon pricing can:

- Incentivize emissions reduction.
- Encourage sustainable investments.
- Support carbon neutrality and net-zero goals.
- Enhance climate risk management.
- Prepare for future carbon regulations and markets.

Internal carbon pricing includes shadow pricing, internal carbon fees, carbon taxes, and cap-and-trade systems.

Remember, internal carbon pricing can drive corporate climate action and support a low-carbon future!

195. Measuring Climate Risk: A Key Strategy for Business Resilience and Sustainability

Climate risk is critical for businesses, encompassing physical, transformation, and liability risks. Decarbonisation and political measures on greenhouse gas neutrality directly impact production, technologies, and product design.

Emissions can be categorized into:

- Direct emissions from company-controlled activities.
- Emissions from power plants supplying electricity.
- Supply chain emissions, including indirect impacts across the entire business ecosystem.

Measuring climate risk helps companies:

- Identify potential impacts on operations and supply chains.
- Assess financial risks and opportunities.
- Inform strategic decision-making.
- Enhance resilience and adaptation.
- Support climate risk disclosure and reporting.

Effective climate risk management is essential for mitigating climate change and enabling sustainable business practices.

196. Measuring Your Company's Carbon Footprint: A Step Towards Sustainable Business Practices

Reducing carbon footprints is essential for combating climate change. To manage them effectively, businesses must first measure their greenhouse gas emissions. Establishing a business-as-usual benchmark allows for targeted reduction efforts.

Many tools and consultancies are available, such as the Carbon Disclosure Project (CDP), which helps businesses transparently measure and report their emissions. Certified agencies can assist in calculating and analysing emissions, enabling companies to identify the highest contributors and implement solutions to reduce them.

Measuring carbon footprint mitigates climate change by:

- Identifying high-emission areas.
- Setting and achieving reduction targets.
- Enhancing transparency and sustainable decision-making.
- Supporting carbon offsetting and supply chain management.

Methods for measuring carbon footprint include:

- GHG Protocol

- Life Cycle Assessment (LCA)
- Carb on calculators
- Supply chain assessments
- Third-party audits

Benefits include:

- Lower emissions and climate impact.
- Cost savings through efficiency.
- Enhanced reputation and trust.
- Compliance with carbon regulations.
- Improved innovation and competitiveness.

Remember, measuring your carbon footprint is a crucial first step towards a sustainable, low-carbon future!

197. Reducing Energy Consumption: Corporate Decarbonization and Climate Action

At COP26, India set the ambitious goal of becoming a net-zero emitter by 2070—a crucial step in addressing climate change. Despite having low per-capita emissions (1.8 tons CO_2), India is the third-largest emitter globally, with 2.9 gigatons of carbon dioxide equivalent (GtCO2e) released annually. A significant portion of these emissions—about 70 percent—is driven by sectors like power, steel, automotive, aviation, cement, and agriculture.

Adopting decarbonization technologies and levers is essential for mitigating climate change. These include renewable energy sources (Solar, Wind, Hydro), energy efficiency measures (LED Lighting, Smart Buildings), electrification (Electric Vehicles, Heat Pumps), and Carbon Capture, Utilization, and Storage (CCUS). Additionally, incorporating strategies such as supply chain optimization, sustainable procurement, and carbon pricing can further enhance efforts.

Implementing decarbonization brings multiple benefits, such as reducing greenhouse gas emissions, improving brand reputation, lowering costs, enhancing competitiveness, ensuring regulatory compliance, and attracting talent.

A clear roadmap—setting science-based targets, conducting energy audits, and investing in technologies—is crucial for successful implementation. By adopting these measures, corporations can reduce climate risks, promote sustainable development, and contribute to mitigating climate change.

198. Adopting Decarbonization Technologies and Strategies for a Sustainable Future

At COP26, India set the ambitious goal of becoming a net-zero emitter by 2070—a crucial step in addressing climate change. Despite having low per-capita emissions (1.8 tons CO2), India is the third-largest emitter globally, with 2.9 gigatons of carbon dioxide equivalent (GtCO2e) released annually. A significant portion of these emissions—about 70 percent—is driven by sectors like power, steel, automotive, aviation, cement, and agriculture.

Adopting decarbonisation technologies and levers is essential for mitigating climate change. These include renewable energy sources (Solar, Wind, Hydro), energy efficiency measures (LED Lighting, Smart Buildings), electrification (Electric Vehicles, Heat Pumps), and Carbon Capture, Utilization, and Storage (CCUS). Additionally, incorporating strategies such as supply chain optimization, sustainable procurement, and carbon pricing can further enhance efforts.

Implementing decarbonization brings multiple benefits, such as reducing greenhouse gas emissions, improving brand reputation, lowering costs, enhancing competitiveness, ensuring regulatory compliance, and attracting talent. A clear roadmap—setting science-based targets, conducting energy audits, and investing in technologies—is crucial for successful implementation.

By adopting these measures, corporations can reduce climate risks, promote sustainable development, and contribute to mitigating climate change.

199. Encouraging Adherence to Office Timings for Sustainability and Productivity

In India, non-adherence to office hours is a significant issue. While employees may arrive on time, many stay late, disrupting their work-life balance, increasing stress, and reducing productivity. Staying late also contributes to higher energy consumption, as air conditioners, lights, and computers remain operational for just a few employees, increasing electricity usage unnecessarily.

Most companies have well-defined policies, processes, and norms for office timings, requiring employees to complete their tasks within the designated hours. However, some employees fail to work efficiently throughout the day due to distractions like social media, excessive phone use, and non-work-related conversations. As a result, they delay tasks until the end of the day, extending their working hours and contributing to greater energy consumption, ultimately impacting the environment.

As per company policy, HR and administration teams are responsible for ensuring offices operate strictly within set working hours, whether for 8-hour or 9-hour shifts. This practice reduces carbon emissions and promotes a healthier work-life balance. When employees leave work on time, they can spend quality time with their families, recharge, and return to work the next day feeling refreshed and productive.

Benefits of Adhering to Office Timings for Climate Mitigation

- Reduced energy consumption: Optimizing energy use during regular hours lowers electricity usage.
- Lower emissions: Limited overtime reduces energy-related emissions.
- Increased productivity: Efficient work during regular hours minimizes the need for extended hours.
- Better work-life balance: Employees experience less stress and better personal time management.
- Reduced traffic congestion: Regular schedules help distribute traffic more evenly.
- Sustainable transportation: Promotes carpooling, public transport, or cycling to work.
- Alignment with renewable energy: Optimized energy use can align with renewable energy sources.
- Minimized paper waste: Reduced overtime means lower office resource consumption.
- Promotes eco-conscious behaviour: Encourages environmentally friendly practices.
- Supports climate policies: Aligns organizations with climate-friendly initiatives.

By adhering to office timings, companies can significantly lower their carbon footprint, contribute to a low-carbon economy, enhance their reputation, improve employee well-being and productivity, and support sustainable development.

Every small change counts. Strictly following office timings is a simple yet effective way to positively impact the environment and foster a more balanced and efficient workplace.

200. Allowing Employees to Work Close to Home: Enhancing Sustainability and Work-Life Balance

Companies often overlook the importance of aligning employees' work locations with their residential addresses. Typically, employees are deployed based on work requirements, with little consideration for their location preferences. While companies may initially factor in an employee's preferred location during hiring, operational demands frequently override these preferences. This leads to employees traveling extensively, whether within the city or to other cities.

For intra-city commutes, employees rely on local transportation or personal vehicles, while inter-city postings often require them to relocate or undertake long, frequent commutes. Those who relocate often return home on a weekly or monthly basis. These travel patterns contribute significantly to greenhouse gas emissions from transportation and conveyance.

Companies can significantly reduce carbon emissions associated with daily commutes by assigning employees to offices closer to their homes. In addition to the environmental benefits, this approach enhances employees' work-life balance, reducing travel-related stress and improving overall productivity.

201. Work from Home: A Sustainable Practice for Reducing Carbon Emissions and Enhancing Employee Well-being

The work-from-home (WFH) culture gained prominence during the COVID-19 pandemic. The restrictions demonstrated that employees could effectively perform their duties without being physically present in the office. While IT companies had already embraced remote work for years, the pandemic prompted almost all industries to adopt WFH practices. Employees have successfully carried out their responsibilities from home, enjoying the benefits of reduced travel, no commuting stress, and better work-life balance.

This shift improved productivity and led to significant environmental benefits, such as reduced carbon emissions from daily commutes and minimized energy consumption in office spaces. However, with the easing of pandemic restrictions, many companies have called employees back to the office, reigniting road traffic and increasing carbon emissions.

Organizations should consider developing policies that enable WFH wherever feasible, based on the nature of the work. Adopting a WFH culture can reduce environmental impact, contribute to a low-carbon economy, and enhance an organization's reputation. It also improves employee well-being, fosters productivity, and supports sustainable development.

Every small change counts, and embracing WFH is a simple yet impactful way to contribute positively to the environment!

202. Reduce Waste in Your Company

Reducing waste in your company is a key strategy for lowering carbon footprints and promoting sustainability. Simple steps like minimizing disposable items, cutting unnecessary printing, reusing materials, and maintaining equipment can significantly reduce waste. Implementing recycling programs, setting waste reduction targets, and encouraging employee participation can further support this effort. Collaborating with suppliers and investing in circular economy practices also helps minimize waste across the supply chain.

By reducing waste, businesses can lower emissions, save costs, enhance their sustainability reputation, and support a low-carbon, circular economy.

203. Reduce Your Company's Value Chain Emissions

Most company emissions occur outside their operations, including upstream activities like supplier manufacturing and downstream activities like product distribution and usage. Transportation, packaging, and materials handling contribute significantly to carbon emissions.

To reduce value chain emissions, companies can analyze their processes and adopt sustainable solutions, such as eco-friendly procurement, product redesign, and collaboration with suppliers and consumers.

Key Strategies to Reduce Value Chain Emissions:

- Optimize Supply Chains: Streamline logistics to cut transportation emissions and encourage sustainable supplier practices.
- Sustainable Sourcing: Use recycled or sustainably sourced materials.
- Energy Efficiency: Adopt renewable energy and energy-efficient operations.
- Product Redesign: Focus on sustainability with recyclable and minimal packaging.
- Reduce Waste: Implement recycling programs and encourage suppliers to do the same.
- Stakeholder Collaboration: Work with customers, suppliers, and partners to promote sustainable practices.
- Science-Based Targets: Set clear emissions reduction goals aligned with climate science.

Reducing value chain emissions helps mitigate climate change, enhances brand reputation, fosters stakeholder trust, and promotes sustainable growth.

Every reduction in emissions counts—collective action can lead to significant environmental improvements.

204. Integrate Climate into Your Company's Business Strategy

Many businesses adopt eco-friendly production and supply chain management practices to achieve net-zero carbon emissions, but more organizations must follow suit. By incorporating climate considerations into their strategies, companies can significantly reduce emissions and support global climate goals, including the Paris Agreement's target to limit temperature rise to 1.5°C.

Key Strategies for Climate Integration:

- Risk Management: Address climate-related risks in operations and supply chains.
- Innovation: Develop low-carbon technologies and sustainable products.
- Resource Efficiency: Reduce energy use, waste, and resource consumption.
- Supply Chain Collaboration: Work with suppliers to adopt sustainable practices.
- Carbon Pricing: Use internal carbon pricing for informed decision-making.
- Stakeholder Engagement: Communicate climate goals with investors, customers, and employees.
- Transparent Reporting: Regularly disclose climate-related risks and progress.
- Leadership Commitment: Ensure CEO and board-level backing for climate initiatives.
- Climate Integration benefited from competitive

advantage, cost savings, enhanced brand reputation, access to sustainable markets, and resilience to climate disruptions, and contribution to global emissions reduction.

Integrating climate into business strategy helps companies mitigate risks, seize new opportunities, foster innovation, and contribute to a sustainable future.

205. Spread Climate Change Awareness and Drive Sustainable Action

While governments raise awareness about climate change, businesses can play a critical role by leveraging their networks, including employees, consumers, and suppliers, to amplify these efforts. By actively promoting climate awareness, companies can influence large groups and contribute significantly to reducing carbon emissions.

How Companies Can Spread Awareness:

- Educational Campaigns: Share information on climate change causes, impacts, and solutions through public campaigns.
- Social Media: Use platforms to promote sustainable practices and engage audiences.
- Employee Ambassadors: Train employees to advocate for climate awareness within their communities.
- Collaborations: Partner with NGOs, environmental groups, and influencers to expand the reach of climate messages.
- Workshops and Training: Offer climate education sessions for employees, customers, and suppliers.
- Climate Literacy: Develop programs to educate stakeholders on sustainability and eco-friendly practices.

Promoting Sustainable Practices:

- Encourage carbon offsetting and offer eco-friendly products.

- Implement energy efficiency and sustainable supply chain practices.
- Sponsor research and share findings to build public understanding.

Companies can benefit from enhanced reputation, greater customer loyalty, cost savings, new market opportunities, and alignment with emerging regulations.

By fostering climate change awareness and adopting sustainable practices, companies can inspire individual and collective action, drive global emissions reductions, and contribute to a sustainable future.

Together, businesses can lead the way toward a climate-conscious society!

206. Create Climate Awareness Among Employees, Clients, and Stakeholders

India is already experiencing the impacts of climate change, including rising temperatures, extreme weather events, and pollution. Yet, awareness about these issues remains low. Companies can play a pivotal role in addressing this gap by fostering climate awareness within their networks.

Key Actions:

- Employees: Educate on climate impacts and company sustainability goals through town halls, in-house contests, and regular updates via official emails. Encourage eco-friendly practices and innovation.
- Clients: Highlight climate-friendly products and services, promote sustainable behaviors, and collaborate on green initiatives.
- Stakeholders: Share transparent progress reports, participate in climate-focused events, and work together on industry-wide sustainability goals.

Benefits:

- Informed decision-making and collective action.
- Innovation through collaboration.
- Enhanced reputation and compliance with climate regulations.
- Resilience against climate-related risks.
- Contribution to global emissions reduction efforts.
- By fostering climate awareness, companies can drive sustainable change, inspire positive action, and help create a resilient, low-carbon future. Awareness is the first step toward a sustainable tomorrow!

207. Optimize Employee Transportation for Sustainability

After the energy sector, transportation contributes major greenhouse gases to the environment. People coming to work via local conveyance or public transport cause emissions.

Companies may encourage their employees to use public transport and carpool with colleagues living nearby. Employers may give some discounts to people who commute by public transportation. This will help significantly reduce carbon emissions and achieve net-zero emissions.

Companies can play a pivotal role in reducing this impact by encouraging employees to adopt sustainable commuting practices. Initiatives such as carpooling with colleagues in nearby areas, promoting public transportation through subsidies or discounts, and fostering a cycling-friendly environment with amenities like bike storage and showers can make a substantial difference. Additionally, offering telecommuting options, transitioning to electric or hybrid company vehicles, and exploring alternative transport modes like electric scooters can further reduce emissions.

Companies can also calculate and offset the carbon footprint of employee travel and incentivize eco-friendly commuting through rewards and recognition programs.

Regularly monitoring and refining these strategies ensures ongoing progress. These measures help achieve net-zero emission goals and bring benefits such as cost savings, improved employee satisfaction, enhanced corporate reputation, and compliance with environmental regulations.

Businesses can significantly contribute to climate change mitigation by optimizing employee transportation and fostering a healthier, more sustainable workplace.

208. Adopting Greener Infrastructure and Equipment: A Path to Sustainability and Reduced Environmental Impact

Companies can significantly reduce their environmental impact by adopting greener infrastructure and equipment. This includes replacing traditional office equipment like printers, computers, and air conditioners with energy-efficient models and transitioning to electric or hybrid vehicles for their fleets.

Investing in renewable energy systems such as solar or wind, constructing sustainable buildings with eco-friendly materials, and implementing smart grid technologies enhances energy efficiency.

Water conservation measures, waste reduction programs, and carbon capture technologies can further support sustainability goals.

By embracing these initiatives, companies lower carbon emissions, improve operational efficiency, strengthen their environmental reputation, and contribute to a healthier planet.

Take action today for a sustainable future!

209. Choosing Sustainable Suppliers: Strengthening Supply Chains and Supporting a Greener Future

Choosing sustainable suppliers is essential for maintaining a company's environmental standards and promoting sustainability. Companies are responsible for partnering with suppliers demonstrating strong environmental practices, such as holding recognized sustainability certifications.

By selecting environmentally conscious suppliers, businesses can reduce their carbon footprint, improve supply chain resilience, enhance brand reputation, comply with regulations, and achieve cost savings.

Key criteria for sustainable suppliers include implementing environmental management systems, using renewable energy, improving energy efficiency, practicing water conservation, reducing waste, sourcing sustainable materials, utilizing low-carbon transportation, and having clear climate policies. To integrate sustainability, companies should assess suppliers, set clear standards, engage them in climate initiatives, monitor performance, and collaborate with industry stakeholders. Regular reviews and training for suppliers ensure continuous improvement.

By prioritizing sustainable suppliers, businesses can drive meaningful change in their supply chains, contribute to global emissions reduction, and support a sustainable future.

Take action today to create a positive environmental impact!

210. Promoting Sustainable Practices Through Green IT Solutions

Encouraging green IT solutions is vital for reducing technology's environmental impact and promoting sustainability. Green IT includes energy-efficient hardware, sustainable software practices, and eco-friendly disposal methods to minimize carbon footprints.

Adopting energy-efficient devices, optimizing data center operations, and utilizing cloud-based solutions help businesses and individuals cut energy consumption and emissions. Additionally, responsible recycling and disposal of electronic waste prevent harmful materials from polluting the environment.

Embracing green IT conserves resources, leads to cost savings, improved efficiency, and more substantial corporate social responsibility, and helps create a more sustainable world.

211. Advancing Sustainability in E-Commerce Practices

Supporting eco-friendly e-commerce practices is also essential for mitigating climate change. E-commerce companies can reduce their environmental impact by optimizing supply chains to minimize emissions, using energy-efficient warehouses powered by renewable energy, and transitioning to electric or low-emission delivery fleets. Other key steps include reducing packaging waste by using biodegradable, recyclable, or reusable materials and offering carbon-neutral shipping options. Promoting sustainable products and partnering with eco-conscious brands also helps build consumer trust.

By integrating sustainability into their business models, e-commerce companies contribute to climate change mitigation and attract environmentally conscious customers, enhancing brand reputation and long-term profitability.

212. Promoting Green Transportation Solutions for Logistics

Adopting green transportation for logistics is essential for businesses to mitigate climate change and reduce their environmental footprint. The logistics sector, heavily reliant on fossil fuel-powered vehicles, contributes significantly to greenhouse gas emissions. Transitioning to greener alternatives, such as electric vehicles (EVs), hydrogen-powered trucks, and biofuels, can drastically reduce emissions and improve air quality. Businesses can further enhance sustainability by optimizing route planning through advanced software to reduce fuel consumption and delivery times.

Investing in multimodal transportation systems that combine rail, shipping, and road transport improves efficiency. Last-mile delivery solutions, such as cargo bikes or electric vans, especially in urban areas, can help reduce congestion and emissions. Governments can support this transition with incentives like subsidies for electric vehicles, tax breaks, and infrastructure for charging stations.

By adopting green transportation practices, businesses support climate goals and achieve cost savings, boost efficiency, and enhance their environmental reputation, creating a cleaner, greener future.

213. Adopting Green Practices in Data Centres for Sustainability

Implementing green data centre practices is crucial for mitigating climate change, as data centres consume vast amounts of energy and contribute significantly to carbon emissions. One key strategy is adopting energy-efficient technologies, such as advanced cooling systems like liquid or free-air cooling, which reduce the need for energy-intensive air conditioning. Transitioning to renewable energy sources, such as solar, wind, or hydroelectric power, further cuts carbon footprints.

Another essential practice is optimizing server utilization by consolidating workloads and reducing underused servers. Virtualization technologies can streamline operations and minimize hardware requirements. Efficient power management systems, including high-efficiency power supplies and energy management software, help monitor and reduce energy consumption. Regularly upgrading to energy-efficient equipment and recycling old hardware also helps reduce waste.

By adopting these green practices, data centres can significantly lower their environmental impact, supporting global efforts to combat climate change.

214. Promoting Eco-Conscious Product Development for a Sustainable Future

Supporting eco-conscious product development is essential for reducing environmental impact and mitigating climate change. This involves designing sustainable products focusing on resource efficiency, longevity, and recyclability. Key practices include using renewable, biodegradable, or recycled materials to minimize resource depletion and waste. Incorporating energy-efficient technologies and low-carbon production processes reduces the carbon footprint of manufacturing and use.

Eco-conscious products should be durable,

extending their lifespan and reducing the need for frequent replacements. Adopting circular economy principles, such as designing for easy disassembly and implementing take-back programs for recycling, helps reduce landfill waste. Collaborating with suppliers who prioritize sustainability further strengthens these efforts.

By integrating these practices, businesses mitigate climate change and meet the growing demand for environmentally responsible products, benefiting both the planet and their bottom line.

215. Enhance Your Office with Indoor Plants: Boosting Productivity and Sustainability

Humans are very close to nature and have a weakness towards it. Nature gives peace, relieves stress, and creates a positive environment, which results in increased productivity.

While artificial decorations like paintings and fake plants add visual appeal, natural indoor plants provide a tranquil atmosphere and numerous benefits for employees and the environment.

Indoor plants purify the air by absorbing carbon dioxide and releasing oxygen, improving indoor air quality. They help regulate office temperature and humidity, support biodiversity, and promote a healthier and more sustainable workplace. Additionally, they enhance mental health, reduce stress, boost productivity, and elevate employee well-being. Natural greenery also adds aesthetic appeal, creating a welcoming and harmonious office ambiance.

Choose the Money Plant, Lucky Bamboo, Snake Plant, Peace Lily, Areca Palm, Aloevera, golden pothos, Warneck Dracaena, Spider Plant, or Zamia ZZ Plant.

Every small step matters—adding greenery to your office can positively impact your workspace and the environment. Bring nature indoors and make your office a healthier, happier place!

216. Transform Your Office Staircase with Indoor Plants: Enhancing Aesthetics and Sustainability

Installing indoor plants in your office staircase enhances the aesthetic appeal and contributes to climate change mitigation and a healthier workspace. Plants purify the air by removing carbon dioxide and toxins while releasing oxygen to improve air quality.

They help regulate temperature and maintain optimal humidity levels, creating a comfortable environment. Adding greenery also boosts employee productivity, enhances well-being, and supports sustainability initiatives.

Low-maintenance plants like Snake, Spider, and ZZ Plants are ideal for staircases, and air-purifying options such as Peace Lily, Dracaena, and Philodendron can be incorporated to add charm. Climbing plants like English ivy or Boston ivy can also be incorporated.

By introducing indoor plants to your office staircase, you can create a visually appealing and environmentally friendly space that promotes health, sustainability, and productivity.

217. Adopt Environmentally Friendly Work Practices: Leading the Way Toward Sustainability

The worsening climate crisis affects everyone, making it essential for individuals and businesses to take steps to reduce carbon emissions. Business activities such as in-person review meetings often require employees to travel long distances, contributing to carbon emissions. To minimize travel, these can be replaced with teleconferencing or videoconferencing. Similarly, reducing paper usage by leveraging digital documentation and email can significantly reduce environmental impact. Careful email targeting to relevant recipients instead of mass emails also conserves energy.

Businesses can adopt various eco-friendly practices to mitigate climate change, such as reducing energy consumption, minimizing waste, promoting recycling, encouraging sustainable transportation, and supporting renewable energy and clean technologies. Additionally, businesses can engage employees in environmental initiatives, implement energy-efficient lighting and equipment, and continuously monitor and improve their sustainability performance.

By adopting environmentally friendly work practices, companies can reduce their carbon footprint, enhance sustainability, improve resource efficiency, and contribute to a low-carbon, resilient future.

Every small change matters; businesses are critical in driving solutions to climate change.

Take action today to make a positive impact!

218. Business Leaders: Catalysts for Climate Action and Mitigation

Business leaders wield significant influence in society in India and globally and are well-positioned to drive impactful climate action. The effects of climate change—from extreme heat and floods to cyclones and pollution—adversely affect businesses by reducing employee productivity and causing health challenges.

Recognizing these impacts, business leaders can address climate change at both personal and business levels, fostering innovation and driving the adoption of high-tech climate technologies to reduce carbon emissions. How Business Leaders Can Mobilize for Climate Action:

- Set Ambitious Goals: Establish science-based climate targets.
- Invest in Clean Technologies: Support renewable energy and green innovations.
- Adopt Sustainable Practices: Implement eco-friendly processes and operations.
- Promote Resilient Supply Chains: Encourage climate-conscious practices across the supply chain.
- Support Research and Advocacy: Invest in climate change research and advocate for effective policies.
- Engage Stakeholders: Educate and involve employees, partners, and customers in sustainability efforts.
- Transparency: Report climate-related risks and opportunities openly.
- Collaborate and lead: Partner with industry associations and embed climate considerations into business strategy.

By mobilizing to combat climate change, business leaders can drive systemic change, support global sustainability goals, enhance their organizations' reputations, and ensure long-term success in a low-carbon and resilient future.

Business leaders play a pivotal role in shaping a sustainable world—every action counts!

219. Encouraging Consumers to Embrace Reusable Bags: A Retail Initiative for Sustainability

With many state governments banning plastic carry bags, the demand for alternatives like paper bags has surged. However, manufacturing paper and plastic bags consumes significant energy and generates carbon emissions, further contributing to environmental degradation.

Additionally, discarded shopping bags create substantial waste that harms ecosystems. While some consumers have started using reusable bags, a significant portion still does not bring their own bags for shopping.

Retail outlets can play a crucial role in promoting reusable bags by offering high-quality, durable options. Although these bags may cost more upfront, they are a sustainable investment, as they can be used repeatedly, reducing waste and carbon emissions over time. The higher cost of reusable bags might encourage consumers to carry their own bags to avoid additional expenses.

How Retail Businesses Can Encourage Reusable Bags:

- Incentives and Discounts: Offer rewards or discounts for customers who bring reusable bags.
- Fees for Single-Use Bags: Charge for single-use bags to discourage their use.
- Affordable Options: Provide durable, affordable reusable bags for purchase.
- Awareness Campaigns: Launch promotions to educate consumers about the benefits of reusable bags.
- Environmental Partnerships: Collaborate with organizations to promote sustainability.
- Recycling Programs: Introduce bag recycling initiatives at retail outlets.
- Employee Advocacy: Train employees to encourage reusable bag use among customers.
- Signage and Promotions: Use eye-catching reminders to highlight reusable bag benefits.
- Loyalty Rewards: Offer points or incentives for customers who consistently use reusable bags.

- Standard Practices: Make reusable bags a core part of the shopping experience.

By fostering these practices, retail outlets can significantly reduce waste, minimize carbon emissions, and inspire consumers to embrace sustainable habits, making a meaningful contribution to mitigating climate change.

220. Minimizing Business Travel: A Sustainable Approach to Reduce Emissions

Business review meetings are often conducted at national, regional, and local levels requiring employees to travel by flight, train, personal car, or public transport. This results in substantial greenhouse gas emissions. The COVID-19 pandemic demonstrated that many of these meetings can effectively be conducted through video or teleconferencing, reducing the need for frequent travel. Prioritizing necessary travel while leveraging virtual collaboration tools can significantly lower carbon emissions and benefit the environment.

Ways to Minimize Business Travel:

- Reduce Transportation Emissions: Limit travel to essential cases to reduce greenhouse gas emissions.
- Encourage Virtual Meetings: Utilize video and tele-conferencing for business reviews and collaboration.
- Promote Sustainable Transportation: Opt for trains, carpooling, or other eco-friendly modes of travel when necessary.
- Implement Green Travel Policies: Establish guidelines prioritizing sustainability in business travel decisions.
- Offset Unavoidable Travel: Invest in carbon offset programs for any essential travel.
- Adopt Remote Training Programs: Develop virtual employee training and development platforms.

- Monitor and Report Emissions: Track and analyze the environmental impact of business travel to identify areas for improvement.

By minimizing business travel, companies can significantly reduce their carbon footprint, save costs, and contribute to a more sustainable and resilient future.

Every step taken toward reducing travel emissions is a step toward combating climate change!

221. Adopt Renewable Energy Sources: A Sustainable Solution for a Low-Carbon Future

As the global climate crisis intensifies, many corporations are transitioning to renewable energy sources to reduce carbon emissions. While some companies have installed solar panels for their energy needs, many have yet to adopt this eco-friendly approach. In India, where coal remains a dominant energy source, companies have immense potential to contribute to sustainability by incorporating renewable energy systems.

Benefits and Actions for Adopting Renewable Energy:

- Reduce Fossil Fuel Dependency: Shift to renewable sources like solar, wind, hydro, and geothermal to lower emissions.
- On-Site Renewable Systems: Install solar panels or wind turbines to power operations.
- Renewable Energy Credits (RECs): Purchase RECs to offset non-renewable energy use.
- 100% Renewable Energy Goals: Commit to powering operations entirely with renewable energy.
- Collaborate with Suppliers: Encourage renewable energy adoption across the supply chain.
- Invest in Innovation: Develop energy storage solutions and support resilient grids.

- Community Projects: Support decentralized, community-based renewable energy initiatives.
- Track and Report Progress: Monitor renewable energy usage and share updates transparently.

By adopting renewable energy sources, companies can significantly lower their environmental footprint, enhance energy efficiency, and lead in combating climate change. Transitioning to clean energy is crucial to a sustainable and low-carbon future!

222. Increase Production of Ecolabel Products: A Path to Sustainable Business Growth

Businesses today are increasingly conscious of environmental protection, integrating climate considerations into their strategies for sustainable development. As consumer awareness about environmental issues grows, the demand for eco-friendly products rises, with many willing to pay a premium for sustainable choices. In both the Indian and global markets, ecolabel products, such as those certified with ENERGY STAR or Eco Logo, are gaining popularity for meeting stringent environmental standards and minimizing environmental impact.

Benefits of Manufacturing Ecolabel Products:
- Promote a Low-Carbon Economy: Support global efforts to reduce emissions.
- Foster Sustainable Development: Align with long-term environmental and economic goals.
- Enhance Brand Reputation: Build trust and loyalty among eco-conscious consumers.
- Ensure Regulatory Compliance: Stay ahead of evolving environmental policies.
- Meet Consumer Preferences: Cater to the growing demand for environmentally friendly options.

- Reduce Risks: Lower environmental liabilities and operational risks.
- Access New Markets: Tap into expanding green product markets.
- Strengthen Supply Chains: Increase resilience through sustainable sourcing and production.
- Contribute to Climate Solutions: Aid in both mitigation and adaptation efforts.
- Lead toward Sustainability: Drive meaningful change for a greener future.

By prioritizing producing eco-friendly products, companies reduce their offerings' environmental impact and position themselves as leaders in the fight against climate change, fostering a sustainable and resilient business model.

223. Focus on Recycled Paper in Tissue Manufacturing: A Greener Approach to Paper Production

Tissue manufacturing is one of the most carbon-intensive paper production processes, emitting more carbon per tonne than other paper types. As the demand for tissue products grows due to the convenience and luxury lifestyle, people often choose tissues over reusable alternatives like handkerchiefs or towels. Producing tissues from virgin paper produces significantly higher carbon emissions than recycled paper.

Although there is a strong market preference for tissues made from virgin paper due to their quality and softness, the environmental impact can be reduced by focusing more on recycled alternatives.

Tissue manufacturing companies can help mitigate climate change by shifting towards recycled paper through various means:

- Conserving Natural Resources: Reducing wood, water, and energy use.
- Lowering Greenhouse Gas Emissions: Minimizing emissions from production and transportation.
- Decreasing Deforestation: Preventing habitat destruction and supporting sustainable forestry.
- Minimizing Waste: Reducing landfill contributions and waste-related pollution.
- Lowering Energy Consumption: Enhancing sustainability by optimizing energy usage.
- Encouraging Closed-Loop Production: Fostering circular economy practices.
- Reducing Carbon Footprint: Lowering the overall environmental impact of tissue products.
- Promoting Sustainable Consumption: Inspiring eco-conscious choices among consumers.
- Enhancing Reputation: Strengthening sustainability leadership and consumer trust.

By prioritizing recycled paper, tissue manufacturing companies can significantly reduce their environmental footprint and contribute to a more sustainable future.

224. Power Generating Companies Can Change the Game: Transitioning to a Greener Future

The Indian energy sector significantly contributes to greenhouse gas emissions, accounting for 75% of the total. According to the Central Electricity Department report, as of June 30, 2024, private companies produced 52.46% of these emissions, state-owned companies 24.13%, and central companies 23.41%.

Out of India's total power generation, coal-based production constitutes 47.28%. Coal, a dirty and polluting energy source, is responsible for 40% of carbon dioxide emissions from fossil fuels. Beyond carbon emissions, coal burning releases pollutants such as mercury, sulfur dioxide, and particulate matter, contributing to respiratory diseases and environmental degradation.

Addressing climate change is challenging without significantly reducing dependence on coal. The primary obstacle lies in the cost of transitioning from coal to renewable energy plants, such as solar, wind, hydro, and biofuel. Despite this, companies face increasing pressure to invest in sustainable energy solutions.

To mitigate emissions, the Indian government has mandated pollution control devices for coal-based power plants. Additionally, under the Paris Agreement, India has committed to reducing its GDP's greenhouse gas emissions intensity by 45% from 2005 levels by 2030. A substantial reduction in emissions from power-generating companies is crucial to achieving this target. Furthermore, India has pledged that 50% of its electricity will be sourced from renewable energy by 2030.

NTPC, India's largest power utility company, is leading efforts by co-firing biomass in its coal-based power plants. Currently using a 7% blend of biomass at its Dadri plant, the company aims to shift 5% of its total generating capacity from coal to biomass. This initiative addresses pollution from crop residue burning by providing farmers with a market for their biomass, reducing emissions, and promoting sustainable farming practices.

Power generating companies can play a pivotal role in mitigating climate change by implementing energy-efficient practices, reducing coal-based production, and increasing renewable energy adoption. Their actions are crucial for lowering global emissions and fostering a sustainable energy future.

225. **Automobile Industries Can Play a Key Role: Driving Towards Sustainable Transportation**

The Indian transport sector contributes 13% of the country's total greenhouse gases, with road transport alone responsible for 90% of these emissions. Automobile industries, including cars, buses, trucks, three-wheelers, and two-wheelers, significantly reduce emissions.

India's commitment to reducing emissions is evident, with Bharat Stage VI (BSVI) norms effective from April 1, 2020—five years ahead of the original proposal for Stage V standards. The Supreme Court's ruling reinforced this, emphasizing that health precedes industry profits.

While BS VI will reduce emissions, particularly in diesel vehicles (up to 70%), petrol vehicles (80% of the automobile market) will reduce emissions by around 25%. To further decrease emissions, automobile companies focus on technologies like Electric Vehicles (EVs), which offer substantial emission reductions.

Electric and solar-powered vehicles can significantly reduce emissions, helping India achieve its climate goals as outlined in the Paris Agreement. The automobile industry's efforts are crucial in promoting sustainable transportation and a healthier future.

226. **Increasing CSR Funds for Environmental Protection: A Step Towards Sustainable Development**

In the face of rising competition, commercial pressures, and evolving regulatory standards, companies are increasingly focusing on sustaining and growing their organizations' wealth. However, environmental consciousness remains relatively low daily, with climate change awareness still limited in India.

Corporate Social Responsibility (CSR) can

help companies meet net-zero emissions targets by integrating social and environmental concerns into business practices.

CSR is a commitment by businesses to address social and environmental issues within their operations. As global environmental challenges intensify, businesses are expected to go beyond financial performance and embed social and environmental concerns into strategic management.

India became the first country in the world to make CSR mandatory in 2014 through an amendment to the Companies Act. Businesses must invest in social areas such as the environment, education, poverty alleviation, gender equality, and more.

Since introducing mandatory CSR in 2014, corporate India has significantly increased its spending. In FY 2022-23, over 24,392 companies registered on the National CSR portal and invested ₹29,986.92 crore across various projects. These projects spanned 14 development sectors, including environmental sustainability, reforestation, renewable energy, health, sanitation, and rural development.

A substantial contribution from corporates is essential to achieving India's national target of 33% forest and tree cover (currently at 24.62% as of 2021).

By allocating more CSR funds toward environmental sustainability, companies can make a significant impact on mitigating climate change through initiatives such as:
- Supporting reforestation and afforestation
- Funding renewable energy and green infrastructure
- Promoting sustainable agriculture and land use
- Supporting climate research and adaptation
- Implementing energy-efficient technologies
- Enhancing water conservation and waste management
- Conserving biodiversity and ecosystem restoration

Increased CSR funding for environmental protection can lead to:

- Reduced carbon footprints
- Enhanced reputation and sustainability leadership
- Compliance with environmental standards
- Support for climate adaptation and mitigation efforts
- Improved stakeholder trust and community engagement
- Access to new markets and opportunities
- Encouragement for innovation in sustainability
- By boosting CSR efforts in environmental protection, companies can drive significant progress toward a sustainable and low-carbon future.

227. The Media: An Ultimate Eye-Opener

Climate change is a pressing global issue, and electronic, print, and social media platforms are vital in raising awareness and shaping agendas. Governments and the public depend on the media to disseminate information and drive development efforts. The 2015 Paris Agreement, ratified by 189 countries, aims to limit global temperature rise to below 2°C. India has committed to net-zero emissions by 2070, with goals like increasing non-fossil fuel energy, reducing emissions intensity, and expanding forest cover.

However, public awareness of these efforts remains low. The media must educate citizens on climate issues—extreme weather, pollution, and resource depletion—while promoting solutions like renewable energy, conservation, and sustainable practices. By targeting state, district, and village levels, the media can inspire action, influence policies, and foster global cooperation, ensuring a sustainable future.

228. Net Zero Banking

The banking and financial sector is becoming increasingly aware of its critical role in facilitating the transition to a low-carbon economy and promoting a sustainable future as the world approaches a pivotal moment in the fight against climate change. **Net Zero Banking,** which aims to align financial services with social and environmental sustainability objectives, is becoming essential for global efforts to combat climate change.

Banks must consider sustainability from multiple angles. They support efforts to decarbonize the economy and finance businesses contributing to carbon emissions. This dual role positions banks as influential catalysts, driving significant changes in global climate statistics.

On a global scale, the market for sustainable finance is projected to grow substantially, with estimates suggesting it could reach USD 23 trillion by 2031, up from USD 3.6 trillion in 2021.

- What is Net Zero Banking?: Net Zero Banking refers to a bank's commitment to achieving a net-zero carbon footprint by balancing the amount of greenhouse gases (GHGs) removed from the atmosphere with the amount released.

- The Need for Net-Zero Financing: To address climate change effectively, the Intergovernmental Panel on Climate Change (IPCC) has recommended limiting global warming to 1.5°C above pre-industrial levels to prevent further environmental damage. However, current temperatures are already 1.1°C beyond these thresholds. As a result, businesses must integrate robust net-zero initiatives into their core operations to safeguard long-term sustainability and adapt to evolving environmental challenges.

- The Rise of Sustainable Banking: The banking industry's dedication to net-zero goals reflects the increasing popularity of sustainable banking practices. In 2023, the United Nations (UN) launched the Net-Zero Banking Alliance (NZBA), comprising 138 of the world's leading banks from 44 countries, managing 41% of global banking assets. These institutions are strategically aligning their operations to achieve net-zero GHG emissions by 2050, which is in line with the Paris Agreement.

Today, banks are aware of shifting stakeholder expectations toward higher sustainability standards. ESG components, particularly the 'environment' and 'sustainability' aspects, are being integrated into internal frameworks, with governance structures evolving to support these initiatives. This reflects the emergence of responsible banking, where environmental impact is increasingly considered alongside traditional profit models. A significant focus is placed on financed emissions resulting from the activities and projects funded by banks.

Banks have been scrutinized for financing activities that contribute to GHG emissions. As societal and regulatory pressures intensify, financial institutions evaluate the environmental impact of the projects they support. By financing green projects, supporting renewable energy, and encouraging energy efficiency, banks are adopting sustainable practices to lower their overall carbon footprint.

Initiatives in India:

In India, Finance Minister Nirmala Sitharaman allocated INR 350 billion (USD 4.2 billion) for the net-zero transition during the Interim Budget 2024. The Securities and Exchange Board of India (SEBI) also developed a framework for sustainable issuance and revised disclosure guidelines for green debt.

Indian commercial banks have supported green finance initiatives for the past decade. These include sustainable agriculture, energy efficiency, and renewable energy projects. The Reserve Bank of India (RBI) introduced a Framework for Acceptance of Green Deposits on April 11, 2023, promoting a sustainable financial ecosystem by encouraging banks and NBFCs to offer green deposits for environmentally sustainable projects. This framework aims to enhance transparency, reduce greenwashing, and support green initiatives.

In 2023, several banks took significant steps, such as:

- Securing a USD 1 billion syndicated social loan, the largest ESG loan in Asia-Pacific.
- Collaborating with the Indian Renewable Energy Development Agency (IREDA) to promote renewable energy projects.
- Launching green deposit policies and lending frameworks aligned with the RBI's Green Deposit Framework.

The Way Forward

The banking sector is central to the sustainable transformation as the world transitions toward a net-zero future. Financial institutions must lead in promoting sustainability, resilience, and environmental responsibility. Net Zero Banking is about minimizing carbon emissions and fostering a future where financial growth and environmental preservation coexist harmoniously.

Banks can drive significant shifts toward a more environmentally sustainable future through sustainable finance products, enhanced transparency, strengthened risk management, and stakeholder involvement. Net Zero

Banking is crucial in mitigating climate change, supporting sustainable development, and fostering innovation while contributing to a low-carbon economy.

229. Corporate Village Adoption: Driving Sustainable Development

Village adoption means working with the community at the grassroots level by understanding ground realities, facilitating and empowering villagers to develop sustainably in their villages, and making the village a modern village with all basic facilities for communities.

Adopting villages under Corporate Social Responsibility (CSR) initiatives can significantly contribute to sustainable rural development and climate change mitigation. Village adoption involves working with communities at the grassroots level to empower and facilitate sustainable growth while ensuring the provision of modern infrastructure like roads, clean water, sanitation, waste management, healthcare, education, and digital facilities.

Benefits of Village Adoption for Climate Action: Corporate involvement can help

- Sustainable agriculture practices.
- Implementation of renewable energy solutions.
- Water conservation and efficient management.
- Reforestation and green initiatives.
- Development of climate-resilient infrastructure.
- Enhanced waste management and recycling efforts.

The government initiated village adoption programs in 2014, urging MPs to lead the charge. However, broader corporate participation is essential to scale the impact, creating model villages that can inspire nationwide replication. Collaborative efforts can ensure sustainable development while addressing climate change and enhancing rural livelihoods.

230. Corporate Adoption of Schools: A Path to Quality Education and Sustainability

In India, government schools face numerous challenges in providing quality education and modern facilities. Many parents prefer private schools offering advanced infrastructure, practical learning opportunities, and access to essential amenities and technology. Unfortunately, government schools often lack these resources, leading to limited educational opportunities for children, particularly in rural areas.

The Role of Corporates in School Development

Corporates have the potential to address these issues by adopting government schools and enhancing their infrastructure through Corporate Social Responsibility (CSR) initiatives. These efforts could include:

- Constructing durable buildings with secure boundaries.
- Establishing smart classrooms with digital learning tools such as projectors.
- Ensuring basic amenities like fans, toilets, sanitation, clean drinking water, and well-maintained roads.
- Installing solar panels to provide sustainable energy.
- Promoting tree plantations and green campuses.
- Conducting teacher training to keep educators updated with modern teaching methods.
- Organizing regular health check-ups for students and maintaining cleanliness.
- Introducing sustainable practices such as composting and effective waste management.

By implementing these measures, corporates can

transform government schools into ideal educational environments, ensuring holistic development for students.

Success Stories: Karnataka's 'Namma Shaale': The 'Namma Shaale' initiative in Karnataka is a prime example of CSR's impact on government schools. Over 60 corporates and non-profit organizations have partnered to improve schools across the state, creating better learning environments and inspiring similar programs nationwide. Broader Benefits: Education and Climate Mitigation

Corporate adoption of schools bridges educational gaps and contributes to environmental sustainability. Initiatives can include:

- Climate education to raise awareness among students.
- Promotion of sustainable practices and eco-friendly habits.
- Development of climate-resilient infrastructure and support for green initiatives.
- Encouraging community engagement in environmental efforts.

Corporate adoption of schools represents a transformative opportunity to improve education, promote sustainability, and empower communities. While some corporates have already taken steps in this direction, a broader commitment is needed to maximize impact. Investing CSR resources in government schools supports education and paves the way for a sustainable and equitable future, nurturing climate-conscious leaders of tomorrow.

231. Felicitation to Employees: Driving Purpose and Sustainability

Attracting Talent through Purpose:

In today's competitive talent market, employees

seek organizations that prioritize sustainability and social impact. Studies reveal:

- 81% of U.S. and 97% of U.K. employers recognize the importance of an environmental strategy for employee satisfaction (WTW HR and Climate Survey).
- Over 60% of workers want employers to address climate change, equality, and poverty.
- Nine in 10 workers prefer meaningful work over higher pay (Harvard Business Review).

Engaging Employees in Climate Action:

- Leaders can cultivate purpose and connection by:
- Listening to employee concerns and measuring environmental awareness.
- Promoting awareness through campaigns and forums for collaboration.
- Incentivizing sustainable behaviours via tailored benefits like clean commute programs, energy subsidies, and volunteer initiatives.

By recognizing employees' eco-conscious efforts, businesses can foster sustainable behaviour, strengthen engagement, and drive both environmental and business success.

232. Eco-Friendly Branch/Unit Recognition Program: Encouraging Sustainable Practices at the Local Level

As the world strives to achieve net-zero carbon emissions by mid-century, with India aiming for 2070, businesses increasingly take voluntary steps to address climate change. Many Indian companies have set ambitious sustainability targets, including Science-Based Targets, renewable energy goals, and carbon pricing.

Recognizing and rewarding branch offices or units for their eco-friendly initiatives can play a key role in reducing

emissions at the local level. How Felicitation Helps:
Recognition Categories:

- Energy Efficiency
- Water Conservation
- Waste Reduction
- Sustainable Transportation
- Eco-Friendly Innovations

Award Ideas:

- Green Branch Award
- Sustainability Champion Trophy
- Eco-Friendly Unit Certificate
- Environmental Leadership Award
- Climate Action Recognition

By recognizing and rewarding eco-friendly initiatives, companies can foster sustainability, boost their reputation, and drive business success.

233. Launch of Additional Climate Fund Schemes by Mutual Fund and Insurance Companies:

Funding is crucial in mitigating climate change through sustainable development across various sectors. Mutual funds and insurance companies are key to raising capital to enhance climate resilience. Several mutual fund companies are already focused on climate funds to address environmental challenges.

Recently, SBI Ventures Ltd., a State Bank of India subsidiary, launched its third climate fund to raise 20 billion rupees. This fund will target small and mid-sized companies working on environmental goals such as waste recycling and emission reduction. The primary focus will be on equity partnerships in these companies.

More climate fund schemes by mutual funds and insurance companies can significantly contribute to climate change mitigation. Such initiatives can increase investments in renewable energy, support sustainable infrastructure, encourage eco-friendly businesses, promote divestment from fossil fuels, and enhance climate resilience.

Climate Fund Scheme Ideas:

- Green Bonds
- Climate Change Funds
- Sustainable Equity Funds
- Environmental, Social, and Governance (ESG) Funds
- Impact Investing Funds

By launching more climate fund schemes, mutual funds and insurance companies can play a pivotal role in supporting climate change mitigation, promoting sustainable development, strengthening their reputation, driving business growth, and contributing to a low-carbon economy.

234. Energy Efficiency and Optimization: Key Strategies for a Sustainable Future

Improving energy efficiency in manufacturing processes is one of the most effective ways to reduce carbon emissions. This can be achieved by upgrading to energy-efficient machinery, adopting smart energy management systems, and optimizing production workflows to minimize waste and lower energy consumption. These measures help reduce carbon footprints and result in significant cost savings, creating a win-win situation for manufacturers and the environment.

Energy efficiency and optimization are essential strategies for mitigating climate change. Key energy efficiency measures include building insulation and retrofitting, installing LED and smart lighting systems, using energy-efficient appliances and equipment, upgrading HVAC (heating, ventilation, and air conditioning) systems, and enhancing windows and doors for better insulation. On the optimization side, energy management systems (EMS), smart grids, and load management systems can be implemented.

By prioritizing energy efficiency and optimization, companies can significantly reduce greenhouse gas emissions, combat climate change, improve energy security, stimulate economic growth, and contribute to building a more sustainable future.

235. Adopting Sustainable Materials and Procurement Practices for a Greener Future

Incorporating sustainable materials into manufacturing processes allows chemical companies to reduce their carbon footprint. This includes using biodegradable, non-toxic materials and recycled resources, which lowers the need for new raw materials and reduces waste. The shift towards sustainable materials grows as companies recognize their environmental and economic benefits.

Sustainable materials include recycled products, low-carbon cement, recyclable plastics, and natural fibres such as hemp and bamboo. Sustainable procurement practices like supply chain assessments, environmental sourcing policies, and circular economy strategies further enhance sustainability.

Key opportunities exist across industries:

- Construction: Sustainable materials and green infrastructure
- Manufacturing: Sustainable packaging and optimized supply chains
- Textiles: Circular fashion and sustainable fabrics

- Electronics: Conflict-free sourcing and e-waste reduction
- Food and Beverage: Sustainable agriculture and reduced food waste

By adopting these practices, companies can improve resource efficiency, enhance brand reputation, and contribute to climate change mitigation, supporting a greener, more sustainable future.

236. Circular Economy Principles: Transforming Manufacturing for Sustainability

A circular economy is an economic model designed to minimize waste and maximize the use of available resources. Unlike the traditional "linear" economy, where products are created, consumed, and discarded, a circular economy seeks to keep products, materials, and resources in circulation for as long as possible. The aim is to establish a closed-loop system that extends the life cycle of materials through practices such as reuse, repair, recycling, and refurbishment.

Adopting circular economy principles can transform manufacturing practices by reducing waste and maximizing resource use. Chemical producers, for example, can recycle materials, reuse products, and recover energy from waste, ultimately lowering their carbon impact.

Circular economy principles help mitigate climate change by reducing greenhouse gas emissions, conserving resources, and promoting energy efficiency. Key principles include designing for circularity, closed-loop production, waste reduction, and regenerative systems.

The benefits of circular economy practices include reduced carbon footprints, cost savings, resource conservation, job creation, and enhanced brand reputation. Industry applications span across sectors:

- Manufacturing: Product design and supply chain optimization
- Agriculture: Regenerative farming
- Construction: Sustainable materials and waste reduction
- Electronics: E-waste reduction
- Fashion: Circular design and sustainable textiles

By adopting circular economy principles, businesses can drive innovation, foster sustainable growth, and create a more resilient, regenerative economy.

237. Sustainable Design: Emphasizing Disassembly and Reuse for a Greener Future

Designing products for disassembly and reuse is a key strategy for reducing carbon emissions in manufacturing. This approach involves creating goods that can be quickly taken apart at the end of their life cycle, enabling components and materials to be reused or repurposed. Making products easier to repair, refurbish, and recycle helps reduce waste, conserve resources, and lower the carbon footprint.

- Benefits: Reduces waste and emissions, conserves resources, extends product lifespan, enhances recyclability, and supports circular economy principles.
- Design Principles: Modularity, easy disassembly, minimal use of fasteners, recyclable materials, and design for upgradeability.
- Industry Applications: Electronics (smartphones, laptops), furniture (modular designs), automotive (EV battery reuse), construction (modular buildings), and textiles (circular fashion).

Implementing Design for Disassembly and Reuse promotes resource conservation, reduces greenhouse gas emissions, and supports sustainable development for a greener future.

238. Establishing Robust Recycling and Waste Management Systems for a Sustainable Future

Adopting waste reduction strategies in the workplace can help minimize landfill waste. This can be achieved by establishing a comprehensive recycling program, forming a green team for annual waste audits, recycling electronics, transitioning to paperless operations, and eliminating plastic bottles. Encouraging litter-free meals and collaborating with suppliers to use recycled packaging also reduces waste.

Implementing strong recycling and waste management systems is crucial in mitigating climate change.

- Benefits: Reducing greenhouse gas emissions, conserving natural resources, lowering waste disposal costs, creating jobs, and improving public health and environmental quality.
- Recycling Strategies: Single-stream recycling, advanced sorting technologies, closed-loop recycling, biogas capture, and composting programs.
- Waste Management Strategies: Waste reduction, landfill diversion, gas capture, incineration with energy recovery, anaerobic digestion, and zero-waste-to-landfill policies.

By implementing these systems, companies can reduce greenhouse gas emissions, conserve resources, promote sustainable development, and improve resource efficiency.

239. **Developing Closed-Loop Manufacturing Systems: A Pathway to Sustainability**

Closed-loop manufacturing systems focus on recapturing and recycling waste materials from the production process to create new products. This method reduces waste while maximizing resource efficiency, resulting in a significant reduction in carbon emissions. It is a real-world example of the circular economy, demonstrating how manufacturers can operate sustainably while fulfilling the global market's needs.

Developing closed-loop systems is crucial in mitigating climate change by reducing waste and emissions. This is achieved through recycling and reusing materials, minimizing production waste, using renewable energy sources, and adopting circular business models. The benefits include conserving resources, decreasing greenhouse gas emissions, improving resource efficiency, supporting sustainable development, and reducing production costs.

Key industry applications of closed-loop manufacturing include the automotive industry (producing parts and materials in a closed loop), electronics (recycling and reusing components), textiles (circular fashion and fabric recycling), general manufacturing (implementing closed-loop systems), and packaging (using biodegradable and recyclable materials).

By embracing closed-loop manufacturing, companies can support sustainable development, boost resource efficiency, foster innovation, and build a regenerative, resilient economy.

240. **Operational and System Improvements: Driving Sustainable Manufacturing for the Future**

Operational and system improvements are vital in

reducing manufacturers' carbon footprints. Companies can significantly increase efficiency by optimizing production processes, integrating innovative technologies, and enhancing energy management. Strategies like predictive maintenance, energy efficiency measures, and lean manufacturing help reduce waste and resource consumption while lowering carbon emissions.

In addition, advancements such as adopting renewable energy, utilizing smart grids, and automating systems contribute to a more sustainable manufacturing environment. These improvements support decarbonisation efforts, lead to cost savings, and enhance market competitiveness.

Operational improvements include energy efficiency measures, renewable energy integration, water conservation systems, waste reduction and recycling, and sustainable supply chain management. System improvements encompass smart grids, energy storage solutions, electric vehicle infrastructure, green building design, and circular economy business models.

These improvements have substantial benefits. They reduce greenhouse gas emissions, improve energy efficiency, enhance resource conservation, offer cost savings, and increase brand reputation and stakeholder trust.

By implementing such operational and system advancements, companies can foster sustainable development, promote resource efficiency, drive innovation, and build a resilient and regenerative economy.

241. Enabling Factors for Successful Decarbonization: Key Drivers for Achieving Sustainability Goals

Successful decarbonization relies on several key enabling factors. First and foremost, a strong commitment from leadership to integrate sustainability into the core business strategy is essential. This commitment must be backed by investments in green technologies and continuous employee training to cultivate a sustainability-focused culture. Additionally, access to reliable data and analytics plays a crucial role in tracking progress and identifying further opportunities for carbon reduction. Tools like the Oizom AQbot, which offers real-time air quality monitoring and data backup, can provide critical insights into environmental conditions, ensuring safety and supporting sustainability goals.

Collaboration across the entire value chain—including suppliers, customers, and regulatory bodies—strengthens the impact of decarbonization efforts. Additionally, leveraging financial mechanisms and incentives for green investments can significantly accelerate the shift to low-carbon operations. When these enabling factors are combined with operational and system improvements, manufacturers are better positioned to achieve their decarbonization objectives.

Addressing these factors can help companies promote sustainable development, enhance resource efficiency, foster innovation, and contribute to creating a resilient and regenerative economy.

242. Leveraging Production Chain Control Systems to Minimize Waste and Enhance Sustainability

Reducing waste is a critical issue that businesses must prioritize for sustainability. Several strategies can be implemented to minimize waste throughout the production chain:

- Invest in New Technologies: Modern technologies like RFID tags can track product movement within the supply chain. At the same time, predictive analytics can help businesses forecast demand more accurately, reducing overproduction and waste.

- Empower Employees: Employees directly involved in the supply chain often have valuable insights into potential waste reduction opportunities. Encouraging them to suggest improvements and take proactive actions can lead to innovative waste-saving solutions.

- Adopt Circular Economy Practices: The circular economy focuses on keeping products and materials in use for as long as possible and minimizing waste. Businesses that engage in circular economy practices can contribute to reducing waste and building a more sustainable supply chain.

- Promote Transparency: Being transparent about waste practices helps raise awareness, fosters accountability, and encourages others to act. It also allows businesses to track progress and identify areas for further improvement.

By embracing these strategies, businesses can significantly reduce waste, create a more sustainable production chain, and contribute to environmental protection.

243. Optimizing Energy Efficiency with Automatic Doors for Sustainable Buildings

Installing automatic doors in offices, hotels, stores, warehouses, commercial buildings, or hospitals can significantly reduce energy consumption and lower costs. These doors are designed to minimize energy loss by improving indoor air quality, enhancing natural daylight, and optimizing overall energy performance.

Using automatic doors will reduce energy bills and contribute to a more sustainable and eco-friendly future.

244. Reducing Energy Consumption with Efficient Air Conditioning Units for a Sustainable Future

In the past, air conditioners were considered a luxury due to their high cost and energy consumption. However, technological advancements have made them more affordable and accessible. To maximize the benefits and minimize energy use, choosing the right air conditioning unit and using it effectively is essential.

By adopting energy-efficient air conditioning systems, companies can reduce energy consumption, lower emissions, cut costs, improve indoor air quality, and contribute to climate change mitigation.

245. Upgrading to Energy-Efficient Aircon and Fan Motors: Reducing Consumption and Emissions

Upgrading your air conditioning and fan motors to more energy-efficient models can significantly reduce annual energy consumption. Ensure that any replacement motor is compatible with your existing unit for optimal performance. Options such as Electronically Commutated Motors (ECMs), inverter-driven aircon motors, high-efficiency DC motors, energy-efficient induction motors, and smart motor technologies are excellent alternatives.

By replacing outdated motors, companies can reduce energy use, lower emissions, cut costs, enhance performance, and contribute to climate change mitigation.

246. Switch to Laptops for Energy Savings and Enhanced Sustainability

Laptops consume up to 80% less power than desktop computers, making them a more energy-efficient choice. Encouraging employees to bring their own laptops or providing them with one can help reduce overall energy consumption.

Laptops offer several benefits, including significant energy efficiency. They use 80-90% less power than desktops.

Additionally, laptops have longer lifespans, generate less electronic waste, and require fewer materials and less energy for manufacturing, thus lowering their carbon footprint. Their portability also supports remote work, helping to reduce commuting emissions. Regarding energy consumption, desktops use between 250 and 500 watts, while laptops use only 20-50 watts.

By switching to laptops, businesses can reduce energy use, lower emissions, save costs, extend device lifespans, and contribute to climate change mitigation.

247. Go Paperless for a Greener Future

Paper production contributes to deforestation, as it is made from wood, and excessive paper use leads to the cutting down of more trees. Reducing paper usage in office operations helps save trees and lowers energy consumption since printers use less idle power.

By transitioning to digital systems, businesses can streamline operations, reduce costs, and minimize environmental impact.

Moving paperwork and compliance online through cloud storage allows easier document access while reducing paper consumption and energy bills.
Benefits:

- Reduced Deforestation: Saves 15% of the global wood harvest.
- Energy Efficiency: Decreases energy consumption by 20-30%.

- Emissions Reduction: Lower greenhouse gas emissions by 20-30%.
- Water Conservation: Saves 10-20% of water used in paper production.
- Waste Reduction: Decreases paper waste by 70-80%.

Paper Usage Statistics:
- 35% of global forest loss is due to paper production.
- 1 tree = 8,333 sheets of paper.
- 1 ton of paper = 17 trees.

Digital alternatives include electronic documents and signatures, digital invoices and receipts, online banking and bill payments, e-books, digital publications, and cloud storage.

Going paperless, businesses can reduce deforestation, conserve water, lower emissions, decrease waste, and contribute to climate change mitigation.

248. Hold Meetings Virtually to Save Energy and Reduce Emissions

Virtual meetings, which became the norm during the pandemic, offer an effective way to maintain business operations without traveling. While face-to-face meetings have advantages, the significant reductions in consumption and travel-related costs during lockdowns highlight the potential for a more sustainable approach.

By continuing to hold virtual meetings, businesses can save time, reduce energy consumption, and cut travel costs.

Travel-related Emissions:
- Air travel: 75% of business travel emissions
- Car travel: 20% of business travel emissions

Benefits:
- Reduced Greenhouse Gas Emissions (up to 90%)

- Energy Savings (up to 80%)
- Cost Savings (up to 50%)
- Increased productivity
- Enhanced global connectivity

Virtual Meeting Tools:

- Video conferencing software (e.g., Zoom, Skype)
- Online collaboration platforms (e.g., Slack, Microsoft Teams)
- Virtual event platforms (e.g., WebEx, Google Meet)

Simple Actions can include choosing virtual meetings over in-person meetings, using video conferencing tools, scheduling virtual events, encouraging remote work, and investing in virtual collaboration software.

By embracing virtual meetings, businesses can reduce travel emissions, save energy, boost productivity, and contribute to global sustainability efforts.

249. Engage Employees in Energy-Saving Initiatives for Greater Sustainability Impact

Employees are invaluable assets to any organization, and their involvement is essential to achieving energy-saving goals. Engaging employees by fostering an open dialogue, offering incentives, and rewarding their contributions is crucial to successfully implementing energy-saving practices. Organizations can achieve significant positive outcomes by encouraging active participation in energy-saving initiatives.

Benefits:

- Increased energy efficiency
- Reduced greenhouse gas emissions
- Cost savings
- Improved employee engagement
- Enhanced corporate reputation

Strategies to Involve Employees:

- Organize energy-saving competitions
- Appoint employee ambassadors for sustainability
- Implement training and awareness programs
- Set up feedback mechanisms for energy usage
- Offer incentives for energy-saving suggestions
- Incorporate gamification and rewards
- Form employee-led sustainability teams
- Provide training on energy-efficient practices
- Share regular progress updates
- Celebrate successes

By actively involving employees, organizations can reduce energy consumption, lower emissions, save costs, enhance employee engagement, and contribute to climate change mitigation efforts.

250. Importance of Regular Maintenance for Energy Efficiency

Regular maintenance plays a vital role in ensuring the longevity and efficiency of business assets, helping reduce energy consumption and operational costs. Establishing a strong preventive maintenance routine can help companies optimize equipment performance and minimize energy use.

Regular maintenance benefits include improved energy efficiency by optimizing equipment performance, reduced emissions by minimizing greenhouse gas output, resource conservation through waste reduction, extended asset lifespans, and cost savings through decreased energy consumption and fewer repairs.

Key maintenance areas include HVAC systems, electrical equipment, industrial machinery, vehicles, buildings and infrastructure, appliances, and renewable energy systems.

Simple actions like scheduling regular maintenance, monitoring equipment performance, replacing worn-out parts, cleaning and inspecting equipment, and upgrading to energy-efficient models can make a significant difference.

By staying on top of maintenance, businesses can lower energy consumption, cut emissions, conserve resources, and reduce costs while supporting climate change mitigation efforts.

251. Switch Off When Not in Use: Simple Actions for Significant Energy Savings

Energy conservation is crucial for reducing costs and emissions, and employee involvement is key in this effort. Encourage your team to switch off lights and electrical appliances in areas that are not in use or consider investing in occupancy sensors for automated energy savings. Don't forget to turn off outdoor lighting when it's not needed during the day.

When not used, everyday items to switch off include electrical appliances like lights, TVs, computers, printers, air conditioners, refrigerators, and electronic devices such as phones, laptops, tablets, and speakers. Additionally, office equipment like printers, copiers, fax machines, coffee machines, water coolers, industrial machinery, pumps, motors, and lighting systems should all be turned off when not in operation.

By consistently switching off devices that are not in use, companies can significantly reduce energy consumption, lower emissions, save costs, extend the lifespan of equipment, and contribute to climate change mitigation.

252. **Calculating Your ESG Performance: A Sustainable and Strategic Approach to Mitigating Climate Change**

ESG, which stands for Environmental, Social, and Governance, is a widely adopted framework that allows companies and investors to assess an organization's sustainability and ethical impact. When effectively implemented, ESG is a powerful, strategic tool in the global fight against climate change, mainly through its Environmental component.

Environmental (E): Climate Action at the Core

- Carbon reduction through renewable energy and energy efficiency
- Sustainable resource use to minimize waste and conserve water
- Climate risk disclosure to build resilience
- Green innovation in low-carbon technologies

Social (S): Empowering People and Communities

- Employee engagement in sustainability initiatives
- Inclusive transitions for vulnerable communities
- Education and awareness across the value chain

Governance (G): Driving Accountability

- Climate-conscious leadership
- ESG-linked executive incentives
- Transparent reporting to track progress

ESG Impact in Action: While not a simple formula, ESG-driven actions lead to:

- Carbon savings and reduced emissions
- Sustainable investment shifts
- Policy influence and industry leadership

Example: ESG in a Net-Zero Strategy. A company adopting ESG can:

Cut emissions by 40% within five years, offset the remaining carbon through reforestation, engage suppliers in emissions reduction efforts, and report progress transparently.

This holistic, ESG-driven approach delivers a measurable impact—not just for the company but for the planet—by aligning business practices with global climate targets, such as the Paris Agreement.

Government Level

"If the planet were a patient, we would have treated her long ago. You, ladies and gentlemen, have the power to put her on life support, and you must surely start the emergency procedures without further procrastination."

Narendra Modi, Prime Minister of India

253. The Urgent Need for Climate Education

The effects of climate change are becoming increasingly evident, with extreme heat waves, droughts, water scarcity, altered rainfall patterns, frequent cyclones, floods, pollution, overpopulation, and rapid deforestation affecting India and the world. While climate change cannot be repaired overnight, immediate actions can be taken to mitigate its impact. This requires collective action and behavioral changes in our daily lives, which can only be achieved through education.

Climate change education is key to altering mindsets, influencing behavior, and encouraging sustainable practices. Children and students, as future change-makers, play a crucial role in this transformation. Additionally, political ideologies must align to ensure the integration of climate change education across the nation. The focus should now shift from mere climate awareness to tangible climate action.

The government could take several steps to encourage green behaviour among students and citizens, such as:

- Incorporating Climate Change into Core Curriculum: Climate change should be introduced as a core subject in schools, colleges, and universities across all streams, Similar to other essential subjects like English, Science, and History, this would help lay the foundation for long-term climate awareness and action.

- Making Climate Change Education Mandatory: Implementing climate change as a mandatory subject from early education (nursery and primary school) to high school and graduation will instill climate-conscious actions in children and students. This approach would also raise awareness among parents, encouraging them to join their children in climate-positive actions.

- Climate Change Specialization in Higher Education: Universities should offer specialized climate change programs, integrated into undergraduate and doctoral courses.

- Curriculum Content: The subject should address the causes, effects, solutions, and action-oriented learning of climate change, promoting critical thinking, creativity, and problem-solving.

- Inclusive and Accessible Learning: The curriculum should be accessible to all students, incorporate diverse perspectives, and support community engagement.

- Global Collaboration: Encourage international connections and collaborative efforts to foster a global sense of responsibility.

A special National Climate Education Policy, framed by both the central and State Governments, could help India become a global leader in mitigating climate change. This long-term strategy may start by prioritizing climate education, setting the foundation for a healthier and more sustainable lifestyle for all citizens.

Climate education is essential to mitigating climate change because it:

- Increases awareness and understanding of the causes and impacts of climate change.
- Empowers individuals to make informed decisions and take effective action.
- Cultivates climate literacy and critical thinking skills.
- Fosters a sense of responsibility and agency among learners.
- Supports the development of climate-resilient communities.
- Promotes sustainable lifestyles and behaviors.
- Prepares students for climate-related careers and opportunities.
- Aids in climate change adaptation and resilience planning.
- Facilitates intergenerational learning and knowledge transfer.
- Helps nurture a global citizenry equipped to tackle climate challenges.

By prioritizing climate education, we can equip individuals with the tools to combat climate change, build sustainable futures, and create a more resilient world.

254. Accelerating Renewable Energy Investment for a Sustainable Future

Renewable energy is pivotal in addressing climate change, yet tapping into its full potential requires substantial investment. The Indian government has launched several initiatives to drive renewable energy development nationwide, aiming to achieve 500 GW of installed electric capacity from non-fossil sources by 2030. Key initiatives like the National Green Hydrogen Mission, PM-KUSUM, PM Surya Ghar, and PLI schemes for solar PV modules are instrumental in advancing this goal. As per the Commerce Ministry, India will offer investment opportunities worth USD 500 billion in clean energy and other sectors by 2030.

India has made considerable progress in its renewable energy journey, surpassing the 200 GW mark in renewable energy capacity by October 2024, with a total renewable-based electricity generation capacity of 201.45 GW, according to the Central Electricity Authority. This milestone highlights years of efforts to harness India's natural resources, such as solar parks, wind farms, and hydroelectric projects. These initiatives have reduced dependency on fossil fuels, improved energy security, and positioned India as a global clean energy transition leader.

The country currently allows up to 100% foreign direct investment (FDI) for renewable power generation and distribution projects, and the government is on track to meet its renewable energy targets. However, the focus must now shift towards timely implementation. Beyond mitigating climate change, investing in renewable energy will create substantial job opportunities across India, reduce emissions from fossil fuel plants, and foster a healthier environment.

By advancing renewable energy production, India

can lead the global transition from fossil fuels to clean energy, aiming for net-zero emissions by 2070. Key renewable sources include solar, wind, hydro, biomass, and biofuels.

Investment in renewable energy can be accelerated through:

- Government policies and incentives
- Private sector financing
- Public-private partnerships
- Crowdfunding and community initiatives
- Research and development funding
- International cooperation and climate finance
- Green bonds and impact investing
- Carbon pricing and emissions trading
- Renewable portfolio standards and targets
- Education and awareness campaigns

Increasing investment in renewable energy will help accelerate the transition to a low-carbon economy, mitigate climate change impacts, and secure a sustainable future for all.

255. Introducing Tax Rebate Schemes to Boost Investment in Renewable Energy Production

The government has already implemented various tax rebate schemes to encourage manufacturers and consumers to invest in and adopt renewable energy. However, there is still significant potential to attract more participation. To further drive this shift, the government can introduce the following income tax rebate schemes to combat climate change:

- Income Tax Rebate for Renewable Energy Investments: A tax rebate of Rs. 50,000 would be available for investments in renewable energy production companies in addition to the 80C limit of Rs. 1.5 lakh (similar to the National Pension Scheme). This would incentivize individuals to invest in renewable energy companies, providing them with tax benefits while ensuring these companies receive the necessary funds for setting up and producing renewable energy.

- Income tax rebate for electric vehicle (EV) purchases: Offers up to Rs. 1.5 lakh in income tax rebates for loans to purchase electric vehicles, applicable to individuals and corporations. This would encourage more people and businesses to opt for EVs, contributing to reducing carbon emissions from traditional vehicles.
- Income Tax Rebate for Rooftop Solar Installations: A rebate of Rs. 1.5 lakh for loans taken to install rooftop solar panels for residential and commercial use. This would motivate individuals and corporations to install solar panels, reducing reliance on coal-based energy sources and promoting sustainable power for everyday devices like phones, laptops, and inverters.

Expanding tax rebate schemes for renewable energy production and usage would reduce carbon emissions, raise awareness, and drive participation in tackling climate change. These schemes could be designed to:

- Offer upfront rebates for renewable energy investments
- Provide ongoing tax credits for renewable energy production
- Accelerated depreciation for renewable energy assets
- Exempt renewable energy equipment from GST
- Offer property tax exemptions for renewable energy installations
- Provide tax credits for renewable energy research and development

- Support community-based renewable energy projects
- Encourage renewable energy adoption in low-income communities
- Foster international cooperation and climate finance
- Continuously evaluate and improve the effectiveness of tax rebate schemes

By implementing these tax rebate schemes, governments can encourage more significant investment in renewable energy production, helping to drive growth, reduce emissions, and ultimately mitigate the impacts of climate change.

256. Investing in Climate Innovations for Achieving Zero Carbon Emissions

Under the Paris Agreement of December 2015, developed countries committed to sharing knowledge and providing $100 billion annually by 2020 to support developing nations in their climate efforts. Unfortunately, this has not materialized as expected. Despite this, India continues to strive towards meeting its climate goals.

The Government of India has made significant policy amendments and allocated annual budgets to meet its Nationally Determined Contributions (NDCs) outlined in the Paris Agreement, aiming to lead the global climate action movement. However, additional investment in climate innovations is crucial for achieving a net-zero carbon emission target by 2050.

In particular, more investment is needed in clean energy innovations, including electrification of the transport sector (electric vehicles and batteries), bioenergy, cleaner 00-heating, green hydrogen, and carbon capture, utilization, and storage (CCUS). Significant technological advancements are also required in solar, wind power, bioenergy, and other sectors. Government support for research and development (R&D) is essential to reduce costs and improve the operational efficiency of these technologies.

The COVID-19 pandemic has slowed investments in clean energy, which is essential for long-term climate change mitigation. Now is the right time to ramp up R&D on climate technologies to limit global warming to 1.5°C above pre-industrial levels and keep it well below 2°C.

Investing in climate innovations for zero carbon emissions can contribute to climate change mitigation in several impactful ways:

- Development of new clean energy technologies
- Improvements in energy efficiency and storage solutions
- Enhanced carbon capture and utilization capabilities
- Support for sustainable transportation and infrastructure
- Promotion of climate-resilient agricultural and forestry practices
- Encouragement of circular economy models and waste reduction
- Development of climate-resilient water management systems
- Support for climate change adaptation and resilience
- Fostering international cooperation and knowledge sharing
- Creation of jobs and stimulation of economic growth

Climate innovations include advancements in renewable energy technologies, smart grids, energy storage systems, electric vehicles, sustainable transportation, CCUS, climate-resilient infrastructure, sustainable agriculture practices, and water management solutions.

Investing in these innovations can be achieved through various means, such as R&D funding, tax incentives, grants, public-private partnerships, climate innovation hubs, regulatory support, international climate finance, education and training programs, innovation challenges, and continuous monitoring and evaluation of innovation impacts.

By increasing investment in climate innovations, governments can accelerate the transition to a zero-carbon economy and mitigate the impacts of climate change, creating a sustainable and resilient future.

257. Global Carbon Tax: A Cooperative Approach to Mitigate Climate Change

A global carbon tax offers an economic approach to reducing carbon emissions. Developed nations tend to have higher per capita carbon emissions due to their industrialized economies, abundant coal plants, and historical development. In contrast, developing countries have lower emissions, primarily because they have fewer industries and are still in the growth stage. While developed countries are focused on reducing their carbon footprints, they face challenges in shutting down their coal plants. Conversely, developing nations, with less industrialization, may feel they are not bound to curb emissions since their per capita emissions are still lower. These disparities create a dilemma: Rich countries can invest more in carbon reduction, but cannot completely eliminate their coal plants. At the same time, poorer nations struggle to meet basic needs and economic growth without adding more carbon emissions.

A cooperative global solution like a Global Carbon Tax could be the best option. A Global Carbon Reduction Incentive (GCRI), a per-ton carbon levy, would require countries with higher per capita carbon emissions to pay into a global fund. Countries with lower per capita emissions would receive a share of this fund. For example, the United States, with emissions exceeding the global average, would contribute around $36 billion, while Saudi Arabia would pay about $4.6 billion. In contrast, countries like Uganda would receive around $2.1 billion, and India would get $41.6 billion. This fund would support carbon reduction initiatives, renewable energy development, and climate change adaptation in developing countries.

A Global Carbon Tax would incentivize developed nations to reduce emissions through domestic policies while encouraging developing countries to adopt carbon-reducing measures to secure financial support. The payouts from the GCRI could be used for renewable energy projects and other sustainable initiatives, helping developing countries transition to greener economies while addressing the impacts of climate change.

At COP29, an agreement was made to kick-start global carbon credit trading, but India rejected the $300 billion climate finance deal, calling it an "optical illusion." Therefore, India should champion the GCRI model at global forums like COP, as it offers a fair and equitable solution for developed and developing nations.

Implementing a global carbon tax can mitigate climate change in several key ways:
- Provides a financial incentive for reducing emissions
- Encourages the transition to low-carbon technologies

- Generate revenue for climate change mitigation and adaptation
- Promotes global cooperation and coordination
- Levels the playing field for businesses and nations
- Encourages sustainable land use and forestry practices
- Supports climate change research and development
- Fosters international collaboration and knowledge sharing
- Helps achieve global climate goals and agreements
- Reducing carbon leakage and competitiveness concerns

By adopting a global carbon tax, the world can reduce greenhouse gas emissions, foster sustainable development, and mitigate the devastating impacts of climate change.

258. Domestic Carbon Tax: A Key Strategy to Curb Carbon Emissions and Drive Sustainability

A Carbon Tax is a levy imposed on companies that burn fossil fuels, such as coal, oil, gasoline, and natural gas. Burning these carbon-rich fuels in sectors like energy and transportation generates greenhouse gases. Introducing a carbon tax is one of the most cost-effective strategies for significantly reducing carbon emissions.

There is widespread global agreement that carbon taxes can facilitate emission reductions, encourage investment and innovation in clean energy technologies, discourage the use of fossil fuels, contribute to achieving sustainable development goals, and generate revenue to support vulnerable communities. Additionally, carbon taxes can help reduce environmental pollution, improve public health, stimulate job creation, and contribute to the transition to a low-carbon economy.

Domestic carbon taxes should be set at higher rates than global standards to have a meaningful impact. For instance, Canada increased its carbon tax from $20 per tonne of carbon emissions in 2019 to $50 per tonne in 2020.

Once implemented, a carbon tax will lead to higher energy tariffs, incentivizing consumers to use electricity more efficiently to reduce their monthly bills. Similarly, increased petrol, diesel, and natural gas prices will encourage consumers to minimize fuel consumption, reducing emissions.

Implementing an effective carbon tax could play a crucial role in achieving India's revised Nationally Determined Contribution (NDC) commitment to reduce the carbon intensity of its GDP by 45% by 2030 compared to 2005 levels.

By adopting such a policy, the government can make significant progress in curbing carbon emissions and transitioning to a more sustainable energy future.

259. Ending Fossil Fuel Subsidies: A Crucial Step Toward a Sustainable and Clean Energy Future

Subsidies play a significant role in India's economy, particularly in sectors where government support is necessary to meet national goals. Historically, subsidies have been crucial in the energy sector to ensure 100% energy access for all citizens. However, much of India's energy consumption still relies on fossil fuels like coal, oil, and gas, dominating the transport sector.

According to the International Institute for Sustainable Development (IISD), subsidies for oil and gas have seen a substantial increase, growing by over 65% from INR 40,762 crore (USD 6.1 billion) in FY 2017 to INR 67,679 crore (USD 10.07 billion) in FY 2019, largely driven by rising oil prices and increased use of subsidized LPG. Conversely, subsidies for renewable energy (RE) fell by 35%, from INR 15,313 crore (USD 2.3 billion) to INR 9,930 crore (USD 1.5 billion) in FY 2019. Consumption subsidies, particularly at the state level, have also increased, with the underpriced electricity subsidy in India estimated at INR 63,778 crore (USD 9.5 billion). While largely unchanged, coal subsidies remain a significant financial burden, with total coal-related subsidies estimated at INR 15,456 crore (USD 2.3 billion) in FY 2019 and expected to increase due to non-compliance with pollution control standards.

To accelerate India's leadership in the fight against climate change, the country must shift away from fossil fuels and accelerate its transition to renewable energy. Phasing out fossil fuel subsidies and reallocating those funds towards renewable energy infrastructure and support could help create greater societal awareness and encourage the growth of clean energy.

Once subsidies for fossil fuels are eliminated, energy tariffs will likely rise, forcing companies and consumers to seek and adopt renewable energy solutions. This transition will create new job opportunities in the renewable energy sector and contribute to improved air quality.

The government can implement this shift by:

- Enacting policy reforms and legislative changes
- Gradually phasing out fossil fuel subsidies
- Redirecting subsidies towards renewable energy and clean technologies
- Implementing carbon pricing and taxation
- Increasing transparency and accountability
- Encouraging public participation and engagement
- Supporting workers and communities in the transition

- Promoting international cooperation and knowledge sharing
- Continuously monitoring and evaluating progress
- Aligning with global climate agreements and goals

By redirecting subsidies from fossil fuels to renewable energy, India can make significant progress toward a greener and more sustainable future.

260. Facilitating Smooth Subsidy Process for Rooftop Solar Panel Installation: Empowering Households and Promoting Clean Energy

Solar energy is a clean and sustainable power source that helps reduce reliance on fossil fuels. The government is keen on promoting the installation of solar panels on residential rooftops, enabling households to generate their own energy. To encourage this transition, the government offers financial support in the form of subsidies for the installation of solar panels in homes.

The PM Surya Ghar: Muft Bijli Yojana is a newly launched government initiative to provide free electricity to households across India. Officially launched by Prime Minister Narendra Modi on February 15, 2024, the scheme offers subsidies covering up to 40% of the cost of installing solar panels on residential roofs. This initiative is set to benefit around 1 crore households and save the government approximately Rs. 75,000 crore annually in electricity costs. **Suitable Rooftop Solar Plant Capacity for households with subsidy support**

Average Monthly Electricity Consumption (units)	Suitable Rooftop Solar Plant Capacity	Subsidy Support
0-150	1-2 kW	₹ 30,000/- to ₹ 60,000/-
150-300	2-3 kW	₹ 60,000/- to ₹ 78,000/-
> 300	Above 3 kW	₹ 78,000/-

Previously, many consumers faced delays in receiving subsidies, leading to frustration and a lack of enthusiasm for recommending solar panel installation to others. Despite the government's supportive schemes, this has resulted in low awareness of rooftop solar panels. However, the PM Surya Ghar initiative presents an excellent opportunity to overcome these challenges.

With increased awareness efforts, mainly through village-level awareness camps, the scheme can be implemented more effectively, helping reduce carbon emissions and promoting clean energy across the country.

261. Facilitating Smooth Solar Energy Sales to the Electricity Board (Net Metering)

With government support and subsidies, many households have started installing rooftop solar panels for self-use. However, energy consumption is not consistent throughout the day. Electricity consumption spikes when more energy-intensive appliances like air conditioners, fans, TVs, and refrigerators are used. On the other hand, when these appliances are not in use, the surplus solar energy produced can go to waste. Net metering provides an ideal solution, enabling households to ensure a steady 24-hour energy supply while reducing their electricity bills.

Households can adopt an on-grid solar system connected to home inverters and the primary electricity grid. This system allows homes to use solar energy for immediate needs and sell any excess power back to the electricity company by feeding it into the grid.

The process of selling surplus solar energy back to the electricity board is known as Net Metering. For example, if a household's on-grid solar system generates

250 units of electricity per month and uses 100 units for their consumption, the remaining 150 units are exported to the electricity board. If households import 50 units from the grid to meet their energy needs, they effectively export 100 units (150 units minus the 50 imported). These 100 units are credited to the household's account, reducing their next electricity bill.

The government can encourage more households and commercial establishments to install on-grid solar systems by promoting net metering. Simplifying and streamlining the net metering process will help achieve the country's solar energy production targets, reduce emissions, and significantly mitigate climate change.

262. Promoting Solar Street Lights in Villages, Cities, and Parks for Sustainable Lighting Solutions

Installing solar streetlights in villages, cities, parks, and along roads enhances safety, security, and convenience at night. In many rural areas in India, the power supply remains unreliable, with frequent unscheduled outages, especially during storms or severe weather conditions, leaving people without lighting after sunset. In some remote and tribal areas, the electricity grid is unavailable, and communities rely on kerosene lamps, contributing to environmental pollution.

Most existing streetlights in urban and rural areas are powered by fossil fuels generated by burning coal and other non-renewable resources. This process emits harmful carbon and exacerbates environmental pollution, leading to climate change.

Solar street lights offer an ideal solution. They harness free, renewable solar energy, which is abundant, noiseless, and pollution-free. While the government has initiated the installation of solar street lights in areas without access to the electricity grid, progress has been slow.

To address this, the government should prioritize installing solar-powered street lights and replacing existing fossil fuel-powered lighting systems in cities, rural areas, parks, and roads. This would help reduce carbon emissions while ensuring all streets and public spaces are well-lit and secure.

263. Expanding LED Bulb Distribution for Energy Efficiency and Climate Action

LED bulbs are significantly more efficient than traditional incandescent bulbs, providing up to 90% energy savings. The "Unnat Jyoti by Affordable LEDs for All" (UJALA) program, launched by India's Prime Minister on January 5, 2015, has become the world's largest zero-subsidy domestic lighting initiative. It aims to address issues like high electrification costs and the emissions caused by inefficient lighting. Today, over 36.78 crore LED bulbs have been distributed across the country.

UJALA's success lies in its innovative approach to energy efficiency, making lighting affordable and accessible while reducing energy consumption. The program has successfully reduced the price of LED bulbs from INR 300-350 each to INR 70-80, making them affordable for households across India. By January 2022, the program had saved 47,778 million kWh of energy annually, avoided 9,565 MW of peak demand, and cut CO_2 emissions by 3.86 crore tonnes.

States nationwide have embraced the UJALA program, which helps lower household electricity bills while improving consumers' quality of life and contributing to India's economic growth.

To further reduce carbon emissions and combat climate change, the government should continue to focus on distributing LED bulbs at subsidized prices through initiatives like UJALA or Gram Ujala schemes. This would facilitate the replacement of outdated incandescent and CFL bulbs in every household, supporting more significant energy efficiency and a cleaner environment.

264. Subsidized Solar Lamps for Rural Communities: A Step Toward Sustainable Lighting and Health Improvement

In many rural areas of India, access to electricity is limited, and power cuts are frequent. As a result, people rely on kerosene lamps or rechargeable battery-powered lights to illuminate their homes. Using kerosene not only emits harmful carbon but also contributes to environmental pollution. Additionally, rechargeable lights consume electricity often produced by burning fossil fuels, releasing carbon emissions contributing to climate change.

To address this issue, the government could distribute solar lamps at subsidized prices in rural areas, helping to reduce carbon emissions. In addition to benefiting the environment, solar lamps would improve the health and well-being of individuals, particularly students who use kerosene lamps for evening study. By making solar lamps more accessible, the government could support both environmental protection and the health of rural communities.

265. Harnessing Solar Power Over National Highways: A Cost-Effective Solution for Sustainable Energy and Infrastructure

As the population grows, the demand for land to build infrastructure such as residential and commercial buildings, schools, factories, and roads increases. Land is also needed for agriculture, further driving up land prices. Solar projects typically require large land areas, and the high land cost adds extra charges to the production tariffs of solar energy.

A cost-effective solution to this challenge is to install solar panels over national highways, where land costs are essentially zero. India's national highway network spans 146,145 kilometers as of 2023. By constructing rooftop solar panels above these highways, there would be no land cost, resulting in more affordable solar energy production.

This innovative approach would generate solar power and offer benefits for vehicles. For example, the shade the solar panel structure provides would reduce energy consumption for vehicle air conditioning. Additionally, shaded roads could reduce wear and tear, leading to fewer road repairs and longer vehicle tire life. The structure could be built at a height of 7 to 10 meters to ensure that it does not obstruct traffic, making it a practical and sustainable solution for energy production and infrastructure maintenance.

266. Dual Benefits of Installing Solar Panels on Canal Tops: A Sustainable Solution for Energy and Water Conservation

India has an extensive network of canals, many of which are connected to rivers. During the summer months, many of these canals experience significant water evaporation. Installing solar panels over these canals could provide dual benefits. First, the required land installation would be free, significantly reducing costs associated with land acquisition for solar energy production. Second, covering the canals with solar panels would help reduce water evaporation, conserving valuable water resources that could be used for agriculture during the dry season.

In April 2012, Gujarat launched the Canal Solar Power Project, using a 532-kilometer stretch of Narmada canals for solar energy generation. This project began with a 1-megawatt installation along a 750-meter stretch and has since been expanded. The cost of generating solar energy in this setup is lower than that of traditional solar plants because land costs are eliminated.

Gujarat's canal network includes about 458 kilometers of main canals, totaling approximately 19,000 kilometers, including sub-canals. Once the Sardar Sarovar Narmada Nigam Limited (SSNNL) canal network is fully completed, it will extend to around 85,000 kilometers. Suppose just 10% of this canal network is used for solar panel installations. In that case, it is estimated that around 2,200 MW of solar energy could be generated, conserving about 11,000 acres of land and saving roughly 20 billion liters of water annually in Gujarat alone.

If this innovative model is adopted across India, it would significantly contribute to achieving the country's renewable energy goals and help India progress towards becoming a net-zero emissions country by 2050.

267. Harnessing Urban Space: Installation of Solar Panels on City Drains for Clean Energy

As urban populations continue to grow in India, approximately 36% of the population now resides in urban areas, consuming large amounts of electricity to power appliances like air conditioners, refrigerators, washing machines, water heaters, induction cookers, grinders, ovens, and electric kettles, with 24-hour electricity access in major cities. However, finding space for solar panel installations in crowded cities is challenging. Every town has an extensive network of covered or uncovered drains that could be leveraged for solar energy generation.

By installing solar panels over these city drains, we could harness solar energy at a lower cost, as the land for such infrastructure would be free. This initiative could raise awareness among city residents about the importance of solar energy in addressing the ongoing environmental crisis and encourage them to install rooftop solar panels at home.

The Delhi government has already begun experimenting with solar panel installations over drains to generate clean energy. Other state governments may consider adopting this approach, though it would require significant investment.

268. Innovative Land Use: Installing Solar Panels on Agricultural Land for Dual Benefits

As the demand for solar energy increases, farmers may consider installing solar panels on unused portions of their farmland where crops are not growing. By setting up solar panels a few feet above the ground, the space underneath can still be utilized for activities like grazing livestock or growing shade-tolerant crops such as nursery plants, vegetables, and small fruit trees. This dual-purpose land use allows farmers to generate solar power while continuing agricultural activities.

This approach offers farmers dual benefits, but the main challenge lies in raising awareness and ensuring effective implementation.

269. **Installation of Solar Panels on Rooftops of Government Schools, Colleges, and Universities**

Schools, colleges, and universities play a pivotal role in shaping students' minds, who, in turn, apply their knowledge to benefit society. As the world faces rising temperatures and climate change, renewable energy is a crucial solution, with solar energy being one of the most accessible and cost-effective options. India aims to achieve 450 gigawatts of renewable energy capacity and 50% of its energy from renewable sources by 2030.

Installing solar panels on the rooftops of government educational institutions would offer multiple benefits. It would allow these institutions to meet their daily energy needs through self-generated solar power, raise awareness among students about the importance of green energy, and encourage them to advocate for solar installations at home. Additionally, it could attract students to eco-friendly campuses, contributing to climate change mitigation efforts.

While several state governments in India have already started installing solar panels in schools and colleges, the pace of implementation is slow. To accelerate progress, policymakers at all levels should prioritize solar panel installations in schools, colleges, and universities, fostering awareness and environmental responsibility.

By adopting solar energy on educational institution rooftops, governments can promote sustainability, reduce carbon emissions, and create a cleaner, greener environment for future generations.

270. **Harnessing the Power of Water: Expanding Hydropower for a Greener Future**

The Government of India is increasing its hydropower

capacity to reduce emissions and boost non-fossil fuel energy. In January 2024, the Ministry of Environment, Forest, and Climate Change (MoEFCC) approved 11.98 GW of pumped storage hydropower projects, the largest batch ever approved. By 2031-32, India aims to grow its hydropower capacity from 42 GW to 67 GW.

Building more dams can help mitigate climate change by:

- Renewable Energy: Hydropower is a clean, renewable source.
- Emissions Reduction: It replaces fossil fuels and lowers emissions.
- Energy Security: Dams provide reliable energy.
- Job Creation: Dam projects generate local employment.
- Flood Control & Water Management: Dams manage water and prevent floods.
- Irrigation Support: Dams provide water for agriculture.
- Carbon Sequestration: Reservoirs store carbon dioxide.
- Climate Adaptation: Hydropower helps communities adjust to climate change.
- Improved Air Quality: Less reliance on fossil fuels improves air quality.
- Energy Independence: Dams reduce dependence on imported energy.

Investing in hydropower projects can reduce emissions, promote renewable energy, and support sustainable development while tackling climate change.

271. Winds of Change: Harnessing Wind Energy for a Sustainable Future

Wind energy is a renewable resource with low operational costs, as wind is a free and abundant fuel source. Once a wind turbine is installed, the primary expense is maintenance, and the turbine does not need to be connected to the grid to function. This makes wind energy more cost-effective and reduces the carbon footprint compared to other forms of renewable energy.

Government support for wind energy plays a key role in mitigating climate change by:

- Providing incentives for the development of wind farms
- Enforcing policies and regulations that encourage wind energy adoption
- Allocating funding for research and technological innovation
- Simplifying permitting processes for wind farm construction
- Fostering public-private partnerships in wind energy projects
- Setting renewable energy targets and standards
- Offering tax credits or grants to support wind energy production
- Advancing grid modernization and infrastructure
- Promoting community engagement and gaining public acceptance
- Developing workforce training programs for wind energy technicians

By supporting wind energy, governments can drive sustainable development, reduce greenhouse gas emissions, and contribute to the global fight against climate change.

272. Bioenergy Revolution: Unlocking Sustainable Business Potential

Bioenergy, derived from organic materials like plants and algae, is crucial in transitioning to a low-carbon energy future. It generates heat, electricity, and liquid fuels, offering a renewable alternative to fossil fuels.

Bioenergy is produced from biomass, which absorbs carbon dioxide during photosynthesis. When converted to energy, biomass releases carbon into the atmosphere, making it a near-zero-emission fuel. Biomass sources include plants, agricultural waste, sewage, and algae.

The three main methods of bioenergy production are:

- Burning: Combustible biomass generates steam to drive turbines.
- Bacterial Decay: Biomass energy is released through bacterial processes.
- Conversion to Fuel: Biomass is turned into biofuels like ethanol and biodiesel.

Bioenergy is the largest renewable energy source globally, representing 55% of renewable energy and over 6% of the total energy supply.

Investing in bioenergy can help businesses drive sustainability while contributing to environmental and economic progress.

273. Holding Electricity Officials Accountable: Combating Power Theft for Sustainable Energy Use

Electricity theft is a significant issue in India, where people tamper with meters, tap cables, misuse service connections, and steal power from transformers. According to the World Bank, power theft costs India 1.5% of its GDP. Recent surveys show that 40% of electricity bills go unpaid, and about a quarter of the total electricity generated is either lost or stolen.

The lack of strict monitoring by electricity officials at all levels allows widespread theft. In both villages and cities, people misuse electricity to cook with electric heaters and run air conditioners, refrigerators, TVs, fans, washing machines, and other appliances without paying for it. Since this power is free through theft, consumers have little incentive to conserve energy. If people were required to pay for the electricity they used, regardless of theft, energy consumption would decrease significantly.

Often, field electricity officials are aware of theft, such as bypassed meters or other methods of stealing electricity, but fail to take action due to corruption or bribes. If officials conducted regular raids and enforced strict penalties, it would raise awareness and discourage theft, leading to more responsible energy use and lower bills. Reduced electricity usage would also result in less fossil fuel consumption.

To tackle this issue, electricity officials must be held accountable for theft within their jurisdictions. The central and state governments should consider implementing policies that ensure officials are responsible for preventing electricity theft and enforcing necessary actions.

274. Boosting Electric Vehicle Adoption: More Investment and Subsidies for a Sustainable Future

Electric vehicles (EVs) play a crucial role in reducing carbon emissions and curbing urban pollution, making them a key strategy in the fight against climate change. The Indian government has recognized this potential and is investing in EV adoption through subsidy schemes. The PM E-DRIVE Scheme, launched with an allocation of Rs 10,900 crore, focuses on accelerating the adoption of EVs, expanding charging infrastructure, and fostering EV manufacturing in India.

This initiative will run from October 2024 to March

2026, with subsidies for electric two-wheelers based on battery power, encouraging consumers to shift to cleaner transportation options.

By providing financial incentives, the scheme helps reduce dependence on fossil fuels and promotes sustainable mobility solutions. With a growing number of EV models from brands like Ola, TVS, and Ather, these efforts can significantly reduce transportation-related emissions, helping India meet its climate goals and improve air quality.

275. Revitalizing the Roads: Implementation of Vehicle Scrappage Policy for Cleaner Air

The Indian government has introduced the Vehicle Scrapping Policy to phase out old, polluting vehicles and replace them with cleaner, more efficient models. The program establishes a network of Registered Vehicle Scrapping Facilities (RVSFs) and Automated Testing Stations (ATSs), with over 60 RVSFs and 75 ATSs currently operational nationwide.

To promote this initiative, vehicle manufacturers offer limited-time discounts for scrapping old vehicles in exchange for Scrappage Certificates. These incentives include up to 3% off the price of new commercial vehicles and 1.5% off passenger vehicles.

This policy aims to reduce vehicular air pollution by 25-30%, promoting cleaner air and greater vehicle efficiency.

Effective implementation of this policy by the central and state governments will significantly reduce carbon emissions and improve public health.

276. Driving Clean Mobility: Effective Implementation of Green Tax for Pollution Control

Green Tax, or pollution tax, is levied on goods and

vehicles that contribute to environmental pollution. The central government has proposed a green tax on petrol and diesel cars older than 15 years and commercial vehicles over 8 years old to encourage using newer, less polluting vehicles. The tax, ranging from 10% to 25% of the road tax, will be applied when renewing fitness certificates.

Electric vehicles, strong hybrids, LPG, and CNG-powered vehicles are exempt from this tax. The generated revenue will combat pollution and fund emission monitoring facilities.

Several states, including Maharashtra, Punjab, and Karnataka, have already implemented similar taxes. All states are encouraged to adopt this policy to help reduce air pollution and encourage cleaner transport options.

277. Advancing Clean Coal: Encouraging the Upgradation of Existing Coal Plants for Sustainable Energy

Coal remains a critical energy source in India, accounting for nearly 70% of electricity generation and fueling industries like steel and cement. However, outdated coal plants without pollution control technologies are major contributors to air pollution and climate change, emitting harmful substances like sulfur dioxide (SO_2), nitrogen oxides (NO_x), and particulate matter (PM).

The government has launched initiatives like the Coal Gasification Mission to gasify 100 million tonnes of coal by 2030. Financial assistance schemes with ₹8,500 crores allocated for coal gasification projects are also in place to promote clean coal technologies.

Upgrading existing coal plants with cleaner technologies will reduce pollution, improve energy efficiency, and help India meet its climate targets under the Paris Agreement. It will also create jobs, stimulate local economies, and enhance public health. Policymakers must prioritize the adoption of clean coal technologies to achieve these goals.

278. Banning Single-Use Plastics: A Step Towards Reducing Pollution and Promoting Sustainability

Due to their convenience and widespread use, single-use plastics, such as plastic bags, cups, straws, and cutlery, have become a major environmental issue. To combat plastic pollution, the Ministry of Environment, Forests, and Climate Change amended the Plastic Waste Management Rules, 2016, and implemented a ban on various single-use plastic items effective July 2022. These include items with high littering potential like plastic earbuds, plates, and thermocol decorations.

Additionally, plastic bags thinner than 75 microns were banned in 2021, and those less than 120 were banned by December 2022. A national task force and state-level committees are working on enforcing this ban and managing plastic waste.

Banning single-use plastics can help reduce plastic pollution, decrease greenhouse gas emissions, conserve resources, and protect biodiversity. It also encourages sustainable consumption and the development of eco-friendly alternatives. Effective implementation of this policy is crucial to mitigating climate change and protecting the environment.

279. Effective Waste Management: A Solution for Health, Sanitation, and Climate Change Mitigation

India faces significant challenges in managing solid and liquid waste, which directly impacts health, sanitation, and the environment. In urban areas, waste collection and recycling face difficulties due to inadequate infrastructure and incomplete household waste collection. Rural areas suffer from a lack of local authorities collecting waste, resulting in open dumping.

The Ministry of Environment, Forest, and Climate Change has introduced waste management rules, including Solid Waste Management (2016) and Plastic Waste Management (2016), and implemented Extended Producer Responsibility (EPR) for managing plastic, e-waste, and other waste types. Additionally, the Swachh Bharat Mission Urban 2.0, launched in 2021, focuses on creating "Garbage-Free Cities" through waste segregation, recycling, and scientific waste management, with a financial allocation of Rs. 1.41 lakh crore over five years.

These policies aim to reduce waste, lower greenhouse gas emissions, and improve sanitation. Effective ground-level implementation is crucial to achieving these goals and mitigating climate change.

280. Incentivizing Recycling: Financial Support for Sustainable Waste Management

India generates over 62 million tons of municipal solid waste annually, with a large portion remaining untreated and causing environmental and health issues. While the government provides funding support under the MSME scheme for waste management innovation, the number of recycling plants remains insufficient.

To address this, the government has introduced guidelines offering financial support for setting up recycling plants, particularly for plastic waste and abandoned fishing gear in coastal areas. Financial assistance includes Rs. 38 lakhs for plastic waste recycling plants and Rs. 48 lakhs for nylon fishing gear recycling plants. The cost is shared between the government (40%) and the project proponent (60%).

More financial assistance and incentives are needed to encourage the establishment of recycling plants in urban and rural areas, which would help manage waste effectively.

281. Swachh Bharat Mission: Transforming Rural Sanitation for a Cleaner India

The Swachh Bharat Mission, launched by the Government of India on October 2, 2014, aims to address the pressing issues of sanitation and waste management across the country, ensuring better hygiene for all. Its main objective was to provide every rural household with a toilet by 2019, working towards an open-defecation-free India.

This initiative has successfully altered the mindset of rural communities, encouraging them to build toilets at home. What started as a campaign has now evolved into a mass movement, with most rural residents no longer favoring open defecation. However, some individuals continue the practice due to long-standing habits.

The Swachh Bharat Mission has brought about significant behavioral change and increased rural populations' awareness of the importance of cleanliness and sanitation. To further strengthen this initiative, Central and State governments must focus on effective implementation, aiming for a completely open-defecation-free India.

This will improve waste management and support the United Nations' Sustainable Development Goals, reducing greenhouse gas emissions and benefiting the environment.

282. Climate Panchayat: An Urgent Need for Grassroots Climate Action

The Climate Panchayat is a village-level initiative to tackle climate-related challenges and promote sustainable community practices. The respected SAMBAD Group, Odisha, recently introduced this innovative concept through the "Parivesh Panchayat" initiative. It aims to foster environmental stewardship, raise awareness on global climate issues, and address local concerns by involving elected representatives from 147 constituencies in Odisha.

By bringing together local MLAs, engaged citizens, and climate advocates, the Parivesh Panchayat encourages meaningful discussions and collective action. This initiative emphasizes the significance of community involvement and inclusive dialogue in developing sustainable climate solutions for Odisha's future.

Policymakers should consider implementing similar Climate Panchayats across the country to help mitigate climate change.

At the grassroots level, the Climate Panchayat can raise awareness, promote renewable energy, encourage sustainable farming practices, enhance water management, and support climate-resilient infrastructure—crucial in creating long-term, community-driven climate solutions.

283. Encouraging More Climate Conclaves for Grassroots Awareness and Action

A climate conclave is a platform where experts, policymakers, and stakeholders come together to discuss climate change and devise strategies to tackle it. Some notable examples include:

- India Climate Policy and Business Conclave: An annual event organized by FICCI and the Ministry of

Environment, Forest, and Climate Change (MoEFCC), this conclave fosters discussions on climate policy, business perspectives, and corporate actions on climate change.

- Climate and Health Solutions (CHS) India Conclave: A collaboration between the Ministry of Health and Family Welfare (MoHFW) and the Asian Development Bank (ADB), this conclave focuses on developing strategies to address climate change's impact on public health.
- National Climate Conclave (NCC): Held in Lucknow, Uttar Pradesh, in 2023, this conclave was organized by MoEFCC, the Department of Environment, Forest, and Climate Change, and the National Centre for Sustainable Coastal Management (NCSCM).
- Youth Climate Conclave: A platform for children and young people to engage on climate change issues, featuring activities like a photography competition, knowledge sessions, and educational excursions.

The government should support and encourage social organizations, corporations, and media houses to host more climate conclaves at the village, panchayat, block, district, and state levels.

These grassroots events will be crucial in spreading awareness about climate change and mobilizing collective action for effective mitigation.

284. Free Distribution of Contraceptives: A Key Strategy for Controlling Population Growth

India's rapid population growth is a significant concern, with a current fertility rate of 1.96 children per woman. This is projected to increase the population to 167 crore by 2050. One effective way to curb this growth is through the use of contraception, especially in rural areas where its usage is still low, leading to unintended pregnancies.

There are various contraceptive options available, such as condoms, oral pills, IUDs, injectable contraceptives, and sterilization. Raising awareness about these methods is essential to preventing unintended pregnancies.

In line with the National Population Policy 2000 and the National Health Policy 2017, the government provides free contraceptives to address family planning needs, including new methods like Injectable contraceptives (Antara Programme) and Centchroman (Chhaya).

This initiative must continue reducing birth rates and achieving the country's sustainable development goals.

285. Implementing a One-Child Policy: A Necessary Step for Population Control and Environmental Sustainability

India is the world's most populous country, with a population of over 140 crore. Due to various social, cultural, and economic factors, it faces significant challenges in controlling its population growth. In contrast, China successfully implemented its one-child policy in 1979, which significantly curbed its population growth. Without a similar approach, India's population may continue to grow uncontrollably.

A National Population Policy advocating for a One-Child Policy is increasingly seen as essential to address this issue. With the support of state governments, the central government must take decisive action to implement such a policy. Population control is crucial in mitigating climate change and reducing environmental pollution.

While the idea of a One-Child Policy is controversial and complex, its ethical, social, and economic implications must be considered carefully. Any policy must be thoughtfully designed and accompanied by support for the communities it impacts.

286. Empowering Girl Child through Education: A Key to Combating Climate Change

Climate change disproportionately impacts women, who face more significant vulnerabilities due to inequality, especially in education. Without access to knowledge and skills, women are less equipped to handle climate-related disasters, resulting in social, economic, and health challenges.

Promoting education for girls can mitigate climate change in several ways:

It enhances girls' green skills, improving their resilience and ability to adapt to climate impacts.

Educated girls gain financial independence, leadership roles, and decision-making power in environmental matters.

It enables girls to make informed choices about marriage and family planning, helping to control population growth and reduce fertility rates.

Education empowers girls to take action on sustainability, climate change, and environmental conservation.

It fosters gender equality and economic growth and reduces reliance on natural resources, addressing the root causes of vulnerability to climate change.

287. Building Sustainable Roads for Pedestrians and Cyclists: A Step Toward Greener Cities

Indian cities are increasingly congested, leading to

daily traffic jams and limited space for pedestrians and cyclists. Widening roads to allocate more space for walking and cycling could reduce vehicle numbers, cut carbon emissions, and improve fitness.

In August 2020, Delhi's Chief Minister, Arvind Kejriwal, announced a plan to redesign 500 km of roads, prioritizing pedestrians and non-motorized transport. Inspired by European cities like the Netherlands, this initiative aims to widen footpaths, create cycling lanes, increase green spaces, and enhance rainwater harvesting. It will ease congestion, improve aesthetics, and ensure safer travel for pedestrians and cyclists.

288. Building Car-Free Cities for a Greener Future

Due to the high number of vehicles on the roads, Indian cities face severe traffic congestion and toxic air pollution. Cities like Delhi often experience dangerous smog, especially in winter. The root cause is excessive vehicular emissions.

Reducing the number of vehicles on city roads is essential for improving air quality. Policy makers could implement car-free weekends or permanent car restrictions in high-pollution areas. This would help reduce carbon emissions and align with the United Nations' sustainable development goals.

By promoting car-free cities, governments can create a healthier, more sustainable, and environmentally friendly urban environment for citizens, contributing to a cleaner and more livable future.

289. Odd-Even Scheme: A Step Towards Cleaner Air

The Odd-Even scheme, first implemented in Delhi in 2016, aims to reduce air pollution by restricting the number of vehicles on the road. Under this scheme, vehicles with odd-numbered registration plates are allowed on odd dates, while even-numbered plates are permitted on even dates.

During high pollution, this traffic rationing method has been successfully used worldwide, including in the US, China, and several European countries. It helps reduce vehicular emissions and improve air quality, especially in cities like Delhi, where air pollution is a serious concern.

State governments can adopt this scheme as an effective measure to curb pollution, reduce emissions, and promote a healthier environment.

290. Expanding Smart Cities: Paving the Way for a Sustainable Future

The Smart Cities Mission, launched in 2015, aims to transform cities into citizen-friendly, sustainable urban hubs. Focusing on essential infrastructure, smart solutions, and better governance, the mission seeks to improve the quality of life by providing clean environments, efficient public transport, and digitalization.

The mission, initially targeting 100 cities, emphasizes energy-efficient technologies, climate-resilient infrastructure, and sustainable development. While progress has been slow, both central and State governments must accelerate the implementation of these projects, expanding the initiative to more cities.

More smart cities will improve urban living and reduce emissions, contributing to climate change mitigation and building a cleaner, healthier environment.

291. Expanding Metro Rail Networks: A Key to Reducing Emissions and Promoting Sustainability

India's metro rail network is growing rapidly, with over 697 km of metro already operational across 15 cities and more under development. Metro systems, which started in Kolkata in 1984, now serve multiple cities like Bengaluru, Chennai, and Delhi, and several others are planning to implement similar projects.

Public transport, especially the metro rail, is far more sustainable than private vehicles. It takes up less road space, consumes less fuel, and emits fewer pollutants. As more people use metros, fewer cars and motorcycles crowd the roads, reducing emissions.

By continuing to expand metro rail networks, cities can promote cleaner, more efficient public transport systems, which can significantly contribute to mitigating climate change and improving air quality.

292. Greening National Highways: Promoting Tree Planting for a Healthier Environment

India's National Highway network spans over 146,145 km as of 2023. In 2015, the government launched the National Green Highways Policy to create green corridors along these highways, promoting tree planting, biodiversity, and carbon sequestration. This initiative aims to reduce carbon emissions, enhance agroforestry, and provide local employment.

The policy allocates 1% of the cost of highway projects annually—around ₹1,000 crore—for tree planting. However, progress has been slow, with only 3,597 km of highways greened by 2020, falling short of initial targets.

It's time for the government to ensure more effective monitoring and execution of this mission. Expanding tree planting along highways will contribute to India's climate goals and improve the environment by increasing carbon sinks, promoting sustainable development, and supporting biodiversity.

293. Promoting Roadside Plantations: Enhancing Greenery on Village and Interconnecting Roads

India boasts the second-largest road network in the world, covering 66.71 lakh km as of 2023. This includes 146,145 km of national highways, 179,535 km of state highways, and 63, 45,403 km of rural, urban, and interconnecting roads. While the forest department and some NGOs are involved in avenue plantations along a few roads, much of the road network remains vacant without greenery.

Contracts are issued each year for new road construction and repairs. A policy similar to the National Green Highway initiative could be applied to these roads, dedicating 1% of the project cost for roadside plantation, beautification, and maintenance. Planting along India's roads would help absorb carbon, support the country's climate goals, engage local communities, and generate employment.

Policymakers may consider implementing such initiatives for environmental and economic benefits.

294. Restoring Coastal Ecosystems: Tree Planting Along India's Coastline

India's coastline stretches over 7,516.6 km, bordering the Arabian Sea, the Bay of Bengal, and the Indian Ocean. This coastal region is home to various ecosystems, including sandy beaches, mangroves, coral reefs, and rocky shores. However, due to frequent weather events, illegal deforestation, and minimal tree density, many coastal forests face degradation, leading to beach erosion, habitat loss, and extinction of wildlife.

Coastal forests play a crucial role in environmental protection, including carbon sequestration. Reforestation and afforestation in these areas can reduce carbon levels, increase the country's forest cover, and help mitigate climate change.

Planting trees along the coastline offers numerous benefits, such as:

- Coastal protection from erosion and landslides
- Carbon sequestration to reduce greenhouse gases
- Protection from storm surges and rising sea levels
- Creation of habitats for wildlife and improvement in water quality
- Prevention of soil salinization and enhanced community resilience

State forest departments may prioritize tree planting along coastal lines to support environmental health and sustainable development.

295. Enhancing River Ecosystems: Tree Plantation Along River Basins

Rivers are vital in India, providing millions with drinking water, irrigation, electricity, and livelihoods. Major cities in India are located along riverbanks, and the country has around 14,500 km of inland navigable waterways. There are twelve major rivers in India, covering an extensive catchment area of about 2.5 lakh square kilometers. These rivers are crucial to the country's geography and economy, flowing from three major watersheds: the Himalayas and Karakoram ranges, the Vindhya and Satpura ranges, and the Western Ghats.

There is significant potential for tree plantation along river basins, which can offer multiple environmental benefits. Planting trees along riverbanks can help in carbon sequestration, reduce soil erosion, maintain biodiversity, and prevent water drainage. Additionally, it can create local employment opportunities.

A dedicated scheme for tree plantation along riverbanks would support India's climate commitments, reduce pollution, and safeguard these crucial water resources.

State governments may consider developing policies to facilitate this important environmental initiative.

296. Promoting Palm Tree Plantations: Strengthening Coastal and River Basin Ecosystems

India's coastline stretches 7,516.6 km, with 2,094 km of island territories and 5,422 km of mainland coast, while its significant rivers cover approximately 14,000 km. Palm trees along coastlines and riverbanks are crucial in stabilizing soil, preventing erosion, and protecting against storms and flooding.

Benefits of Palm Tree Plantation:

- Coastal Protection: Palm roots help stabilize sand, prevent erosion, and protect against cyclones and storms. Their foliage reduces wave energy and shields the shoreline.
- River Basin Protection: Palm trees prevent soil erosion, reduce flood risks, and filter water, improving water quality.
- Climate Mitigation: Palm trees absorb CO_2, reducing greenhouse gas emissions.

In response to increasing lightning strikes, the Odisha government has approved a plan to plant 19 lakh palm trees to mitigate fatalities. Palm trees' natural ability to absorb lightning offers further environmental protection.

Planting palm trees in these areas will help safeguard ecosystems and support climate resilience.

297. Supporting Mangrove Conservation: Protecting Coastal Ecosystems and Enhancing Resilience

Mangrove forests are natural barriers against coastal flooding, storm surges, and rising sea levels. Their complex root systems stabilize coastlines, reducing wave erosion. Planting and preserving mangroves along the coast can protect communities and ecosystems from extreme weather events while providing vital habitats for various species.

Benefits of Mangrove Conservation:

- Coastal Protection: Mangroves reduce the impact of storm waves and tsunamis by slowing water flow and preventing erosion.
- Wildlife Habitat: They offer breeding grounds for sharks, turtles, fish, and invertebrates.
- Enhanced Fish Farming: Mangroves help improve fish farm yields by providing natural food and protecting farms from flooding.
- Boost Crab Production: Crabs rely on mangrove trees for nesting and breeding.

The Union Budget 2023-24 introduced the 'Mangrove Initiative for Shoreline Habitats & Tangible Incomes (MISHTI)' to promote and conserve mangroves for their biological productivity, carbon sequestration, and role as a bioshield. Now, state governments must act in their respective states to expand Mangrove coverage and environmental resilience.

298. The Importance of Urban Plantation: Combatting Pollution for a Healthier Future

Approximately 36% of India's population resides in urban areas, with this number set to increase to 40% by 2030 and 53% by 2050. Unfortunately, many Indian cities face alarming pollution levels, with PM 2.5 exceeding safe limits. According to a Swiss report, 22 of the 30 most polluted cities globally are in India, 14 ranking in the top 15. This pollution surge is linked to the growing urban population, more vehicles on the road, industrial emissions, construction activities, and deforestation for infrastructure. The limited number of trees cannot absorb enough carbon to mitigate the effects.

Urban plantation is a critical solution to combat air pollution. Planting trees along city roads, road dividers, streets, parks, in front of homes, shopping complexes, and commercial buildings can improve air quality.

Local urban bodies and state forest departments should take the initiative to green their cities, providing citizens with healthier, cooler environments. Urban plantations will not only reduce pollution but also help mitigate climate change.

299. Celebrating National Holidays with Tree Plantations: A Step Towards a Greener India

India celebrates key national holidays like Independence Day (15 August), Republic Day (26 January), and Gandhi Jayanti (2 October). These occasions are an opportunity for citizens to reflect on their contributions to the nation, and what better way to honor the country than by protecting the environment?

The government can incorporate environmental protection into these celebrations by promoting tree planting, banning single-use plastic, and raising awareness about reducing carbon emissions. For example, in 2017, Panchkula district in Haryana launched the "Green Your District" program, planting over 5,000 trees and encouraging citizens to participate in environmental activities, like a "selfie with tree" campaign.

Governments may encourage all districts to celebrate national holidays and festivals in eco-friendly ways, such as tree planting initiatives, to:

- Raise awareness about climate change
- Encourage community involvement
- Promote sustainable development
- Support biodiversity
- Inspire individuals to take action for the environment

300. Integrating Tree Plantation into Education: Empowering Future Generations for Environmental Stewardship

Tree planting may be included in the curriculum in schools, colleges, and universities to protect the environment and raise awareness. It should be treated as a subject, integrated from Nursery to Ph.D., with marks awarded for activities like sapling creation, plantation, nurturing, and tree survival and growth. This initiative aligns with the constitutional duty to protect the environment and ensures that future generations contribute to reducing carbon.

Recommended Action Plan:

- Include tree plantation as an independent subject from Nursery to Ph.D.
- Students create saplings from fruit seeds at home and plant them in school or their local community.
- Allocate dedicated class periods to teach tree care and plantation techniques.
- Students update teachers on the progress of their trees through videos or calls.
- Teachers visit plantation sites annually to guide students.
- Students are graded based on the survival and growth of their trees.

- Encourage students to promote environmental protection in the community for extra marks.

This approach will help create a culture of sustainability, empowering students to take actionable steps toward a greener, healthier planet.

301. Ensuring Sapling Survival: The Importance of Taller Plants for Better Success Rates

The survival rate of trees is just as important as the planting itself. Every year, millions of saplings are distributed, often around 1–3ft in height, resulting in a low survival rate due to inadequate growth in nurseries. This has hindered India's progress in reaching the national tree cover target of 33%.

However, saplings that are 5ft or taller have a much higher survival rate for several reasons:

- Reduced animal damage: Taller plants are less vulnerable to pests and animals.
- Increased visibility: Larger plants are easier to monitor and maintain.
- Improved soil conditions: Taller plants have better root systems, allowing them to access more nutrients.
- Better disease resistance: Taller plants are more immune to pests and diseases.
- Enhanced photosynthesis: Larger plants have more leaves for better energy production.
- Increased drought tolerance: Taller plants lose less water through transpiration.

To improve the survival rates of saplings, the government should focus on distributing taller plants (5ft and above) from nurseries for better long-term success.

302. The Need for 3-5 Years of Maintenance in Plantation Projects

Maintaining trees is as important as planting trees. Each year, the government, forest departments, and NGOs plant trees with the help of funding. However, many of these trees fail to survive due to cattle grazing, natural calamities, or road expansion. Currently, most planting projects are only maintained for one year, and if saplings are damaged, no replacement takes place. This undermines the purpose of these efforts.

Saplings need 3 to 5 years of maintenance for proper growth and survival. Some saplings grow within a year, while others need more time. Regular maintenance and replacement are essential for ensuring the long-term survival of trees. A 3-year audit should also be conducted to compare the initial plantation count with the actual survival rate, aiming for at least 75% survival despite challenges like weather and animal interference.

To ensure success, policymakers may include a 3-5 year maintenance and replacement clause in plantation projects, with funds held in retention accounts until an audit is completed. Additionally, a penalty should be imposed if the survival rate falls below 75%. Prioritizing this extended maintenance will significantly improve tree survival rates and the overall success of plantation efforts.

303. Holding Forest Officials Accountable for Survival of Saplings from Government Nurseries

Each year, crores of saplings are produced in government forest nurseries, but the survival rate remains low. As a result, India has struggled to meet its target of achieving 33% green cover of its geographical area over the past five decades.

While the forest department plants a small portion of saplings from the forest nursery at designated sites, many are distributed to the public or organizations as part of various government schemes. The recipients—whether individuals or institutions—must submit documents like Aadhar cards or application forms for plantation, but once the saplings are handed over, forest officials rarely follow up. In many cases, these saplings are planted without proper protection or damaged by cattle, leading to a very low survival rate.

The lack of post-distribution monitoring means that forest officials often have little to no knowledge of how many saplings survive after distribution, despite maintaining records. This oversight is particularly concerning, given taxpayer funds are used to produce these saplings. On the other hand, when the forest department plants trees, officials are held accountable for a survival target of 75%, which drives them to ensure better care and survival of these trees.

Forest officials must be held accountable for the saplings produced in their jurisdiction to improve the survival rate. Policymakers should establish clear guidelines for officials to track and ensure the survival of at least 75% of the saplings produced in their nurseries. This accountability should be monitored at every level, from the Principal Chief Conservator of Forests to the Forest Guards.

Additionally, the number of saplings produced and their survival rate should be publicly displayed each year to ensure transparency and greater responsibility in forest management.

304. Holding Forest Officials Accountable: Tracking and Reducing Unauthorized Tree Cutting

Each year, crores of trees are cut down for infrastructure development projects like road widening, residential buildings, commercial complexes, industrial facilities, schools, and colleges, often without proper authorization. In some cases, individuals also cut trees for personal use without obtaining the necessary permissions. Unfortunately, forest officials typically do not track or maintain records of tree numbers in their jurisdictions, leaving them unaware of the annual increase or decrease in tree population.

If forest officials take legal action against unauthorized tree cutting, it would raise awareness and discourage illegal deforestation. People would be more likely to seek prior approval from the competent authority before cutting trees.

Forest officials should be assigned specific targets for increasing the tree count in their jurisdiction to improve the situation, such as a 10% to 20% annual increase in each village, city, panchayat, block, district, and state.

Additionally, performance appraisals should be based on the year-over-year increase or decrease in tree numbers. This approach will encourage officials to engage more with local communities and proactively increase tree numbers, benefiting the environment and contributing to national reforestation goals.

305. Holding Horticulture Officials Accountable for the Survival of Fruit Trees in Government Schemes

Each year, the Horticulture Department allocates significant taxpayer funds for producing fruit tree saplings like Mango, Coconut, Amla, jamun, and Sapota, as well as other vegetables and flowers. These permanent fruit trees, which will remain in the environment for many years, are distributed to farmers under various schemes, with subsidies provided for their maintenance. However, there is often no clear data on the survival rates of these saplings in specific villages, panchayats, blocks, or districts.

Currently, there is no publicly available information on the survival rates of these trees or the year-over-year increase or decrease in their numbers. To address this, horticulture officials—from field officers to the Director/Commissioner—should be held accountable for the survival of the fruit trees they distribute. They should track how many trees have survived and maintain records of the annual increase or decrease in tree numbers in their respective areas.

This approach would encourage a more careful distribution of subsidies based on actual survival rates rather than the number of allotted saplings. Officials would be more focused on ensuring the survival of the trees they distribute, as they would be directly accountable for their success.

Policymakers should consider creating policies to hold horticulture officials accountable for tree survival under various schemes. Annual survival data should be published for each village, panchayat, block, district, and state. These statistics should be made publicly available, and officials should undergo performance appraisals based on tree survival rates in their jurisdiction.

By implementing such policies, we can motivate officials to engage more with local communities and ensure the success of horticulture nursery projects, ultimately contributing to a more significant number of healthy, thriving trees across the country.

Let's work together to promote sustainable horticulture practices and support long-term environmental health!

306. The Urgency of Wildlife Conservation: Protecting Ecosystems and Biodiversity

Wildlife is an integral part of the ecosystem, maintaining biodiversity and ecological balance. Human activities such as infrastructure development, agricultural expansion, and poaching have led to the extinction of several species and to biodiversity loss. Wildlife plays a vital role in maintaining the food chain and ecosystem stability. For example:

- The decline of one species can disrupt the entire ecosystem, affecting the survival of other species.
- Poaching and habitat destruction lead to the extinction of animals like elephants, tigers, and deer, which future generations may never see.
- All wild animals contribute to the ecosystem, and their loss can negatively impact human resources like food and water.

The Indian government has implemented several measures for wildlife protection, including the Wildlife Protection Act (1972), the establishment of national parks and sanctuaries, and the creation of organizations like the Wildlife Crime Control Bureau and the National Tiger Conservation Authority. Initiatives like E-Surveillance in Kaziranga and special patrolling strategies aim to curb poaching and illegal wildlife trade.

However, despite these efforts, wildlife extinction continues, and urgent action from policymakers is needed to strengthen conservation efforts and ensure the survival of these critical species.

307. Urban Bodies Are Key to Environmental Protection in Cities

Around 36% of India's population lives in urban areas

managed by urban bodies such as Municipal Corporations and City Councils. These bodies are responsible for public health, welfare, safety, infrastructure, and environmental protection, including waste management, sanitation, clean water supply, and urban greenery.

Urban areas face high pollution levels due to population density and lifestyle choices. With many high-profile individuals living in cities, pollution affects everyone, making it crucial for urban bodies to act.

Urban bodies should consider planting 5,000 to 10,000 saplings annually within and around the city for three consecutive years. This initiative can help create a greener, healthier city by:

- Reducing pollution
- Beautifying urban spaces
- Improving public health
- Cooling the city
- Supporting biodiversity

Additionally, urban bodies should enforce stricter penalties for environmental violations to raise awareness and encourage responsible behavior. By prioritizing these actions, urban bodies can enhance public health, combat climate change, and improve the quality of life for residents.

308. Encouraging Farmers to Adopt Sustainable Alternatives to Agricultural Biomass Burning

Agriculture plays a vital role in India's economy, with year-round crop cultivation producing large amounts of agricultural waste, including crop residues. Due to a lack of sustainable waste management practices, farmers often resort to burning these residues, leading to environmental pollution and numerous health issues. Recently, cities like

Delhi have experienced severe air pollution from crop residue burning in neighboring states such as Haryana and Punjab. Despite efforts by both the central and state governments to address this issue through campaigns and promoting sustainable practices like converting crop residues into energy or compost, the problem persists.

The key to solving this issue is effectively implementing sustainable waste management practices. Policies should prioritize the benefits to farmers and be framed collaboratively by both the central and state governments. These policies should focus on education, empowerment, and awareness-building among farmers and provide technical solutions for proper waste management.

The government can significantly contribute to mitigating climate change by facilitating these efforts and raising awareness about the harmful effects of agricultural biomass burning.

309. Implementing a Permanent Ban on Firecracker Production for a Healthier, Greener Future

India is a religious country that celebrates festivals like Diwali, Durga Puja, and Eid by bursting firecrackers, contributing significantly to air and noise pollution. Firecracker use generates greenhouse gases and harmful pollutants, including PM10 particles, which lead to respiratory problems and other health issues. Firecrackers also produce noise levels that can cause hearing loss, sleep disturbances, and high blood pressure. Additionally, they are responsible for numerous injuries, especially to the eyes, hands, and face.

In response to climate change and pollution, some state governments imposed restrictions on firecracker sales during Diwali in 2020, supported by the National

Green Tribunal and Supreme Court rulings. A permanent ban on producing, selling, and using firecrackers is crucial for mitigating pollution, promoting public health, and supporting sustainable practices.

310. Monthly One-Day Break for a Greener Planet

Climate change, extreme weather events, pollution, and health issues all affect us. While many view these as natural processes, human activities contribute to rising temperatures and environmental degradation. The COVID-19 lockdown showed us that reducing human activity can significantly lower pollution levels, with emissions dropping by over 30%.

To address climate change, let's implement a monthly one-day break to reduce emissions and raise environmental awareness. Just as we took months of lockdown for health reasons, a one-day environmental break each month can:

- Lower emissions by halting industrial and transportation activities
- Raise awareness and promote sustainable practices
- Conserve resources like water and energy
- Encourage eco-friendly activities such as tree planting and park cleanups
- Improving air quality and supporting biodiversity
- Foster global cooperation on climate change

Policymakers could consider this monthly break a step towards environmental sustainability, creating a collective effort to protect our planet.

311. Observing Environmental Dates to Boost Public Awareness

India, known for its rich cultural diversity, celebrates

many festivals yearly. These festivals mark various occasions, seasons, religious events, and significant anniversaries. While India has long celebrated many cultural and religious observances, environmental days like World Environment Day, Earth Day, Water Day, Wildlife Day, and VanMahotsav week have also gained prominence in recent years.

However, many more environmental dates can be observed to raise further awareness about critical environmental issues. The Government of India could consider marking additional environmental days at the national level to engage the public and foster greater environmental consciousness.

List of Environment Dates to be celebrated in India:

1. World Environmental Education Day – 26th January
2. World Wetlands Day – 2nd February
3. International Polar Bear Day – 27th February
4. World Wildlife Day – 3rd March
5. International Day of Action for Rivers – 14th March
6. Global Recycle Day – 18th March
7. World Sparrow Day – 20th March
8. International Day of Forests – 21st March
9. World Water Day – 22nd march
10. World Meteorological Day – 23rd March
11. World Health Day – 7th April
12. Earth Day – 22nd April
13. World Day for Safety and Health at Work – 28th April
14. World Migratory Bird Day – 2nd Saturday of May and October
15. International Day of Light – 16th May
16. World Bee Day – 20th May
17. National Endangered Species Day – 3rd Friday of May

18. International Day for Biological Diversity – 22nd May
19. World No-Tobacco Day – 31st May
20. World Bicycle Day – 3rd June
21. World Environment Day – 5th June
22. World Food Safety Day – 7th June
23. World Oceans Day – 8th June
24. World Day to Combat Desertification and Drought – 17th June
25. Van Mahotsav (Forest Festival) - 1st Week of July
26. World Population Day – 11th July
27. International Day for the Conservation of the Mangrove Ecosystem – 26th July
28. World Nature Conservation Day – 28th July
29. International Tiger Day – 29th July
30. International Day of the World's Indigenous People – 9th August
31. World Lion Day – 10th August
32. World Elephant Day – 12th August
33. International Day of Clean Air for blue skies – 7th September
34. International Day for the Preservation of the Ozone Layer – 16th September
35. Zero Emission Day – 21st September
36. World Rhino Day – 22nd September
37. World Tourism Day – 27th September
38. World Habitat Day – 1st Monday of October
39. World Migratory Bird Day – 2nd Saturday of May and October
40. World Animal Day – 4th October
41. International Day for Disaster Reduction – 13th October
42. International Snow Leopard Day – 23rd October
43. International Day of Climate Action – 24th October

44. World Cities Day – 31st October
45. World Tsunami Awareness Day – 5th November
46. World Toilet Day – 19th November
47. World Soil Day – 5th December
48. National Energy Conservation Day – 14th December
49. International Mountain Day – 11th December

312. Upgrading Government Offices to Energy-Efficient Devices for Sustainability

Many government offices across various levels in India still use outdated infrastructure and energy-consuming devices, leading to high electricity consumption and emissions. Upgrading to energy-efficient devices can significantly reduce energy use and environmental impact. Benefits of upgrading include:

- Reduced energy consumption: Energy-efficient devices use less power, lowering emissions.
- Cost savings: Energy-efficient devices reduce electricity bills.
- Lower carbon footprint: Helps decrease greenhouse gas emissions.
- Leadership in sustainability: Government offices can set an example for others.
- Improved air quality: Less energy use reduces fossil fuel dependence and pollution.
- Compliance with environmental regulations: Ensures adherence to green standards.

Government offices can reduce costs, improve sustainability, and contribute positively to the environment by upgrading.

313. Empowering Rural India: Promoting Clean Fuels Through Village-Level Awareness

Around 64% of India's rural population still cooks with traditional fuels like wood and kerosene. Despite the success of the Pradhan Mantri Ujjwala Yojana (PMUY), which provides LPG connections to rural households, many still use wood due to old habits.

Awareness campaigns at the village, panchayat, and block levels by Govt. are crucial to encourage the switch to clean fuels, which can help:

- Reduce health risks from indoor air pollution.
- Lower environmental impact by reducing carbon emissions.
- Save money on fuel costs with affordable alternatives like LPG or biogas.
- Promote community involvement and support from local leaders.

Educating rural populations on the benefits of clean fuels can improve health, reduce emissions, and foster a more sustainable future.

314. Holding Municipality Officials Accountable for Building Bye-Law Violations

Building byelaws regulate critical aspects of construction, such as building size, height, safety, and architectural design. They ensure orderly development and safeguard against fire, earthquake, and structural failures. The 2016 Model Building Byelaws were updated to address environmental concerns, safety, and technological advancements.

When individuals or developers apply for construction permits, the approving authority—whether a Municipality or urban development authority—issues plans based on these by-laws. However, many buildings are not being constructed according to the approved plans, and officials fail to monitor compliance adequately. While some officials visit sites, violations often go unnoticed until construction is nearly complete, making it challenging to correct unauthorized work.

Monitoring construction as per approved plans is just as important as issuing permits. Municipal and building construction officials must be held accountable for ensuring compliance with building bylaws, which promote environmental sustainability and reduce carbon emissions. Key areas of accountability include:

- Failure to enforce energy-efficient standards
- Allowing non-compliant materials
- Neglecting insulation or ventilation requirements
- Overlooking water management systems
- Ignoring design flaws
- Not inspecting construction sites
- Overlooking environmental impact assessments
- Neglecting community concerns
- Lack of transparency in permitting
- Insufficient training and expertise

By holding officials accountable, we can ensure that buildings comply with modern, sustainable bylaws, significantly mitigating the impact on climate change.

315. Holding Municipal Officials Accountable for Effective Waste Management

Municipal solid waste management is a significant environmental concern in Indian cities. Improper waste management leads to pollution and public health risks. With the growing urban population and increased waste generation, many cities dispose of waste in open dumps and landfills, creating further environmental issues.

The Ministry of Environment, Forests and Climate

Change and pollution control boards have set rules for safe waste disposal, but effective implementation is crucial. Municipal Commissioners and officials are responsible for waste segregation, collection, transportation, and disposal. Accountability must be established for failures such as delayed waste collection, open transportation, and improper dumping.

By holding officials accountable, we can ensure proper waste management, leading to a cleaner, healthier, and more sustainable environment.

316. Strengthening Laws for the Preservation and Cleanliness of Water Bodies

India is home to numerous water bodies, such as ponds, lakes, and talab's, essential for maintaining ecological balance. These water bodies provide drinking water, recharge groundwater, control floods, support biodiversity, and sustain the livelihoods of many people.

India is facing a severe water crisis, with millions of people lacking access to clean water. By 2030, the water demand is expected to exceed the available supply, leaving 40% of the population without access to clean drinking water. Neglect and poor conservation of water bodies, including the disposal of sewage, waste, and religious offerings, are major contributors to this crisis.

Although the government has allocated funds to clean water bodies, progress has been slow due to inadequate monitoring at the panchayat, block, and district levels. Urban areas also show little concern for the cleanliness of their water bodies, which harms the environment and poses public health risks.

To address this issue, accountability must be established for maintaining clean water bodies at both rural and urban levels. Policymakers should consider implementing stricter laws to protect and restore these vital resources.

Enforcing stringent laws can help preserve water bodies, combat climate change, and ensure a sustainable future.

317. Restoring Our Beaches and Rivers: A Call for Environmental Action

Oceans and rivers, among the world's most valuable natural resources, have become polluted garbage dumps, posing risks to human and marine health. Each year, billions of pounds of harmful substances—ranging from plastics and trash to sewage, pesticides, and oil—contaminate our coastlines and river basins. Excessive nitrogen and phosphorus from fertilizers and animal waste further degrade these water bodies due to human activities. The damaging effects of pollution on our beaches and rivers are becoming increasingly clear, and urgent action is needed to address the problem.

In 2018, the Union Ministry of Environment launched a campaign to clean 24 beaches, water bodies, and 24 polluted river stretches in preparation for World Environment Day on June 5. The ministry formed 19 teams and allocated ₹10 lakh for each site to help with the cleanup, involving local students and residents in the effort. However, progress has been slow, and more funds and closer monitoring are required to accelerate the drive for cleanliness.

Cleaning our beaches and river basins not only helps reduce greenhouse gases but also ensures the health of both humans and marine life. Let's work together to clean our environment and combat climate change!

318. Enforcing Stricter Laws to Prevent Illegal Tree Cutting for Development

With the rising population, trees are being cut yearly to make space for infrastructure like roads, schools, industries, and parking lots. Many people bypass the required permissions or fail to follow government guidelines, leading to penalties and consequences.

Under the Indian Forest Act, cutting a tree without proper authorization can result in a fine of Rs. 10,000 or up to 3 months in prison. However, in many cases, violators either bribe officials or pay the fine, making the penalty ineffective. The Environment Authority Act mandates tree transplantation or planting new saplings, but this is often ignored due to poor enforcement.

Stronger laws and stricter monitoring are needed to protect trees in urban and rural areas. The government can balance development and environmental protection by implementing stringent laws, safeguarding our future, and mitigating climate change.

319. Enhancing Accountability in Enforcing PUC Certificates for Environmental Protection

With the rise in population, urbanization, and increased income levels, more people are adopting a luxurious lifestyle, leading to a significant increase in vehicles on the road. This surge, particularly in two- and four-wheelers, has caused a sharp decline in air quality, pushing it beyond safe limits.

According to the Central Motor Vehicle Rules, 1989, every vehicle must obtain a Pollution under Control (PUC) certificate to ensure it does not exceed the permissible limits of air pollution. These certificates can be obtained from government-approved centers, and if a vehicle lacks a valid

PUC certificate, the owner faces a fine of Rs. 1,000, with Rs. 2,000 for subsequent offenses. The certificate remains valid for six months, requiring renewal at regular intervals. The Insurance Regulatory and Development Authority of India (IRDAI) also mandates a valid PUC certificate to renew motor insurance policies.

Although the law was initially strictly enforced, many vehicle owners now neglect to renew their PUC certificates. Traffic police, local authorities, and motor vehicle officials are no longer as diligent in checking compliance, resulting in widespread casualness. This negligence undermines the original intent of the PUC system, threatening the environment.

Officers must be more accountable to address this issue, and vehicles must be checked more rigorously and frequently. Policymakers should consider deploying additional resources in motor vehicle departments or Regional Transport Offices (RTOs) to ensure stricter enforcement and better environmental protection from air pollution.

320. Ensuring Accountability in Environmental Law Compliance for Development Projects

The Environmental Impact Assessment (EIA) process, introduced in India in 1994, grants projects like mines, dams, and industries access to land, water, forests, and other resources, provided they receive prior Environmental Clearance. However, despite having strong environmental laws, the real issue lies in the lack of effective monitoring of projects after they've been approved.

In March 2020, the Ministry of Environment amended EIA norms to simplify business processes. While India regularly updates its environmental regulations, many projects fail to comply with approved terms and conditions due to insufficient monitoring by environmental officials.

To address this, officials must closely oversee projects from approval to completion, ensuring that all environmental guidelines are followed. By enforcing these laws and guidelines more strictly, the government can better mitigate climate change and safeguard the environment.

321. Expanding National Green Tribunal Reach: Establishing Benches at the State Level

The National Green Tribunal (NGT) was established on October 18, 2010, to address environmental protection and conservation issues, offering a specialized body for the swift resolution of cases related to environmental disputes. Initially, NGT has benches in five locations: New Delhi (principal), Bhopal, Pune, Kolkata, and Chennai. However, this coverage is insufficient given the growing environmental concerns nationwide.

Setting up NGT benches in each state and union territory ensures faster case resolution and improves public access. This would facilitate quicker case disposal, enhance public confidence, and encourage more individuals to report environmental violations, contributing to better protection of natural resources. Establishing benches at the state and district levels would further strengthen environmental governance and promote sustainable development.

322. Establishing Weekly Environmental Courts in High Court and District Courts for Faster Case Resolution

According to the 2024 National Crime Records Bureau (NCRB) report, many environmental cases remain pending in courts across the country. Environmental degradation continues due to human actions, with violations of environmental laws occurring nationwide, but many cases go unfiled due to a lack of awareness. Currently, such cases are handled in regular courts, with no specialized benches for environmental violations, aside from the NGT, which is in only five locations.

Establishing dedicated weekly courts or benches in high courts and district courts to hear environmental cases could raise public awareness and discourage environmental violations. With more people filing cases in the public interest, this initiative would also speed up case resolution. Policy makers should consider this approach to benefit public health and environmental safety.

Creating separate environmental courts would reinforce the government's commitment to environmental protection and climate change mitigation, ensuring a more sustainable future for all.

323. Streamlining the Process for Filing Environmental Complaints and FIRs

Environmental violations are widespread in India, but most citizens are unaware of how to file complaints or FIRs against violators. Currently, few cases are filed, mainly by NGOs and activists. To address this, the government should make it easier for citizens to report violations through online platforms, dedicated helplines, and an efficient redressal mechanism at the local and panchayat levels.

India has various environmental laws, such as the Air and Water Pollution Act, the Wildlife Protection Act, and the Environmental Protection Act.

By simplifying complaint processes, raising aware-ness, and offering protection for whistleblowers, the government can encourage more people to take action, thus ensuring better enforcement and environmental protection across the country.

324. RBI to Introduce Guidelines for Banks on Climate Finance Lending to Boost Green Infrastructure

The World Economic Forum's 2021 Global Risk Report highlighted climate action failure as a top global risk, projecting a need for $5 trillion annually for green infrastructure, far beyond current public funding commitments. Green Finance, which supports environmentally sustainable projects like renewable energy, clean transportation, and waste management, is crucial for combating climate change.

In India, funding for climate projects is currently insufficient, and banks are slow to invest in green initiatives due to a lack of clear guidelines. To address this, the Reserve Bank of India (RBI) may develop policy guidelines requiring banks and financial institutions to allocate a portion of their lending to climate finance. This would support renewable energy, sustainable agriculture, green transportation, and other projects that help mitigate climate change and reduce greenhouse gas emissions.

325. Expanding Irrigation Infrastructure to Boost Agricultural Productivity

In India, irrigation relies heavily on canals, groundwater, tanks, and rainwater harvesting projects. As of 2013-14, 36.7% of agricultural land was irrigated, with the rest depending on the monsoon. Groundwater accounts for 65% of India's irrigation needs. Currently, 51% of agricultural land is irrigated, with the remainder reliant on rainfall. Inconsistent rainfall can severely impact crop production, limiting farmers to one crop yearly and reducing food output.

Expanding irrigation facilities can enable year-round crop production, helping to sequester more carbon from the atmosphere and contribute to economic growth. The government should prioritize increasing irrigation infrastructure to mitigate climate change and boost the country's GDP.

326. Promote Organic Farming

Climate change increasingly threatens agriculture, with rising temperatures causing heat waves, droughts, and floods. Agriculture contributes 33% of global greenhouse gas emissions, but organic farming can help reduce these emissions, conserve energy, and enhance carbon sequestration.

Organic farming is growing in India, covering 2.78 million hectares as of 2020, but it still represents only 2% of the country's total agricultural land. States like Madhya Pradesh, Rajasthan, and Maharashtra lead the way, but many other states have not fully embraced organic farming.

To reduce greenhouse gas emissions and protect the environment, both central and state governments should focus on expanding organic farming across India.

327. Promoting Organic Compost for Sustainable Farming and Soil Health

As the demand for organic farming grows, organic compost is crucial in improving soil health and reducing the need for chemical fertilizers. Made from decayed carbon-based materials like fruits, vegetables, and manure, it enhances soil fertility.

States like Chhattisgarh and Odisha have already implemented government schemes to promote organic compost, which has improved farmers' incomes, reduced chemical fertilizer use, and boosted soil health.

Central and state governments should expand these initiatives to support organic farming and help mitigate climate change.

328. Encouraging Community Composting in Rural Areas

In rural India, household waste is often discarded openly, contributing to environmental pollution and releasing harmful gases like methane and nitrogen. With 64% of the population residing in rural areas, solid and liquid waste management is crucial to the Swachh Bharat Mission, which aims to improve cleanliness and hygiene.

To tackle this issue, rural households can create their own compost dockyards to turn kitchen, garden, and agricultural waste into compost, reducing the need for chemical fertilizers and mitigating greenhouse gas emissions. States like Odisha are already assisting rural households in setting up these composting systems. Policymakers should implement more such initiatives to promote composting and help combat climate change.

329. Promoting Better Sanitation and Water Access in Rural Areas

The Central Rural Sanitation Program (1986) and Swachh Bharat Mission (2014) have made significant strides in providing sanitation facilities in rural India. However, many rural households still lack proper toilets and drinking water access. While toilets have been built in many villages, issues like the absence of water pipes prevent their use.

Improving sanitation and drinking water access in

rural areas will help reduce open defecation, lower methane emissions, and contribute to climate change mitigation. The government must continue prioritizing sanitation to improve public health and the environment.

330. Promoting Awareness Camps to Eliminate Open Defecation

While the Swachh Bharat Mission has led to the construction of toilets in rural households, many still do not use them due to the lack of water supply. This leads to open defecation, which harms public health and the environment.

To tackle this, district and block administrations should organize awareness camps in panchayats and villages, educating people on the benefits of using toilets. Promoting behavioral change will help reduce open defecation, lower methane emissions, and contribute to mitigating climate change.

331. Promoting Ecolabel Products to Encourage Sustainable Consumption

Ecolabeling is a certification process that identifies products and services with environmentally friendly features. The "Ecomark" scheme, launched by the Government of India in 1991, helps consumers easily identify eco-friendly products based on environmental criteria, from raw material extraction to disposal.

Products like cloth napkins, reusable shopping bags, rechargeable batteries, and LED bulbs are ecolabel items. By increasing awareness of ecolabel products, the government can encourage sustainable consumption, helping to protect the environment and mitigate climate change.

332. Ensuring Accountability of Pollution Control Board Officials for Effective Environmental Compliance

The Central Pollution Control Board (CPCB), established in 1974, is responsible for promoting the cleanliness of water bodies and controlling air pollution in India under the Water (Prevention and Control of Pollution) Act, 1974, and the Air (Prevention and Control of Pollution) Act, 1981. State Pollution Control Boards (SPCBs) are tasked with implementing similar functions at the state level.

Before starting projects such as thermal plants or commercial or residential developments, NOC (No Objection Certificate) or clearances from the respective Pollution Control Boards are required. These clearances come with conditions to ensure compliance with environmental standards. However, the Pollution Control Boards often fail to closely monitor whether projects adhere to these environmental guidelines once approved.

Pollution Control Board officials must be held accountable for the environmental compliance of new and long-standing projects. They should actively monitor key environmental parameters like water quality, air quality, and waste management practices to ensure projects meet the required standards.

Policymakers should establish a framework to hold these officials accountable. This would lead to more effective monitoring and better environmental protection, which would help reduce violations and mitigate climate change.

333. Boosting Rainwater Harvesting

In India, many urban areas face water shortages due to inefficient distribution and limited supply. While city governments source water from various sources, such as rivers, lakes, and groundwater, some regions still lack proper access, and continuous piped water is unavailable.

Rainwater harvesting (RWH) can significantly help address this issue. Collecting and storing rainwater for future use ensures a reliable water supply, especially in areas with limited surface or groundwater resources. RWH systems are already in place in several buildings, societies, and public spaces, with notable examples like Delhi's international airport, which uses over 300 wells to recharge groundwater.

Despite increasing awareness and government initiatives, only 8% of India's rainwater is currently being harvested. To combat water scarcity and reduce vulnerability to climate change, the government should prioritize expanding RWH systems across the country.

334. Support and Recognize Climate Heroes

Climate Heroes are pivotal in combating climate change, promoting sustainability, and safeguarding the environment. They can include:

- Activists: Advocating for climate action and policy change.
- Scientists: Researching and developing innovative climate solutions.
- Journalists: Raising awareness and reporting on environmental issues.
- Entrepreneurs: Creating sustainable products and services.
- Community Leaders: Driving local climate initiatives and education.
- Policymakers: Shaping and enforcing climate policies.

- Educators: Teaching about climate change and sustainability.
- Artists: Using creative platforms to raise environmental awareness.
- Youth Leaders: Motivating young people to engage in climate action.
- Indigenous Leaders: Protecting traditional lands and knowledge.
- Everyday Individuals: Making sustainable choices and inspiring others.
- NGOs: Working on the ground to protect the environment through tree planting, awareness campaigns, and more.

Recognizing and encouraging these Climate Heroes can amplify efforts to protect our planet and foster a sustainable future.

335. Felicitation Programs for School Students to Inspire Climate Action and Leadership

Education begins in school, where the foundation for every individual's growth is laid. Students are the future of our nation, and what they learn in school shapes their contributions to society. Many students strive to excel academically or in extracurricular activities such as essay competitions, debates, sports, music, and dance. Recognizing their achievements through felicitation motivates them to continue their efforts.

With climate change becoming an urgent global concern, schools must encourage students to engage in climate action actively. Government-supported recognition programs can inspire students to contribute to mitigating climate change.

Benefits of Felicitation:

- Inspires students to take action on climate change.
- Fosters environmental awareness and education.
- Encourages sustainable living practices.
- Cultivates future leaders in climate action.
- Aligns with national climate goals.

Felicitation Ideas:

- National Climate Champion Awards
- Environmental Excellence Certificates
- Climate Action Scholarships
- Green School Awards
- Climate Leadership Recognition Programs

Levels of Felicitation: Block Level, District Level, State Level, and National Level

Recognizing and motivating students through these awards can play a crucial role in promoting climate-conscious behavior and shaping the leaders of tomorrow.

336. Felicitation for College Students: Encouraging Innovation and Leadership in Climate Action

Government recognition through felicitation can motivate college students to contribute to climate change mitigation. By rewarding outstanding contributions to environmental sustainability, the government can inspire future leaders to take action.

Felicitation Categories:

- Best Climate-Friendly Project
- Outstanding Environmental Research
- Climate Leader of the Year (Student/Faculty)
- Green Innovation Award
- Sustainability Champion

Felicitation Ideas:

- Awards Ceremonies
- Certificates of Appreciation
- Scholarships for Climate-Related Studies
- Funding for Climate-Focused Research/Projects
- Internships with Climate Organizations

Levels of Felicitation: College Level, University Level, and UGC Level

By recognizing and motivating college students, the government can foster climate action, raise environmental awareness, promote sustainable living, and support national climate goals—ultimately helping to mitigate climate change.

337. Felicitation for Schools: Inspiring Climate Action and Sustainability Initiatives

Government recognition through felicitation can motivate schools to take meaningful steps towards mitigating climate change and promoting sustainability.

Felicitation Categories:

- Best Green School
- Climate Action Award
- Environmental Education Excellence
- Sustainable Infrastructure Award
- Community Engagement Award

Felicitation Ideas:

- Award Ceremonies
- Certificates of Appreciation
- Funding for Environmental Projects
- Recognition on Government Websites/Social Media
- Scholarships for Students/Staff

Levels of Felicitation: Block Level, District Level, State Level, and National Level

By recognizing schools through these awards,

the government can inspire climate action, enhance environmental education, promote sustainable practices, and support national climate goals, ultimately helping to mitigate climate change.

338. Felicitation for Colleges: Motivating Climate Action and Sustainability Leadership

Government recognition through felicitation can motivate colleges to take active steps in addressing climate change and promoting sustainability.

Felicitation Categories:
- Best Green College
- Climate Action Award
- Sustainability Research Excellence
- Environmental Education Leadership
- Community Engagement Award

Felicitation Ideas:
- Award Ceremonies
- Certificates of Appreciation
- Funding for Climate-Focused Research/Projects
- Recognition on Government Websites/Social Media
- Scholarships for Students/Faculty

Levels of Felicitation: University Level and UGC Level

By recognizing colleges through such awards, the government can inspire climate action, raise environmental awareness, promote sustainable practices, support national climate goals, and contribute to mitigating climate change.

339. Monitor Effective Utilization of Eco Club Funds

The National Green Corps (NGC), initiated by the Ministry of Environment and Forests in 2001, aims to sensitize schoolchildren about environmental issues through eco-clubs. Over 1 lakh schools are involved in the program, and the Centre for Environmental Education (CEE) supports its implementation.

A State Steering Committee oversees the program at the state level, with the Principal Secretary of the Forest and Environment Department as Chairman. The Centre for Environmental Studies (CES) acts as the state nodal agency, coordinating training and reporting.

To ensure the effective utilization of funds, the government should closely monitor the activities and outcomes of these eco-clubs, including training Master Trainers and eco-club teachers and proper reporting from district committees. This will help ensure the funds are used effectively to promote environmental education and conservation.

340. Felicitation for Media Houses: Encouraging Effective Climate Reporting and Awareness

Recognizing media houses for their climate reporting is an effective way for governments to foster climate awareness and action. By honoring media outlets for their coverage, governments can encourage comprehensive and widespread reporting on climate issues, essential for addressing climate change.

Strong climate reporting helps raise public awareness, motivates individual action, and influences policy decisions. Governments can support this by providing resources and incentives for media organizations to produce quality climate-focused content.

Some possible government initiatives to promote climate reporting include:

- Awards and Recognition: Establish awards or recognition for outstanding climate journalism.

- Training and Resources: Offering training, workshops, and access to climate experts for journalists.
- Grants and Funding: Providing grants or financial support for climate-related reporting projects.

Government support for media coverage of climate issues can yield several benefits:

- Enhanced public awareness and engagement
- Improved accuracy and depth in climate reporting
- Increased credibility of media outlets focusing on climate topics
- Promotion of solution-oriented journalism

By supporting climate reporting, governments can help build a more informed public and drive urgency for climate action.

341. Government to Boost Climate Fund Schemes for Sustainable Growth and Net Zero Targets

Climate change is projected to impact India's economy significantly, potentially leading to an annual GDP loss of 3% to 10% by 2100. In a business-as-usual scenario, India could see a GDP per capita loss of 2.6% by 2030, 6.7% by 2050, and 16.9% by 2100.

The effects of climate change are expected to be widespread, affecting key sectors such as health, agriculture, labor productivity, and infrastructure. Once a certain warming threshold is crossed, these impacts may become irreversible, potentially leading to ecological, social, and economic collapse. Therefore, limiting global warming to 1.5°C and transitioning to a low-carbon economy in the coming decades is essential. However, achieving this target will require substantial investments -7 % to 18% of India's 2019 GDP. To meet India's Nationally Determined Contribution (NDC) targets, an estimated annual investment of USD 167 billion from 2016-2030, or around 8% of India's GDP in 2015, is needed.

Meeting the 1.5°C target will require even higher investments, ranging from 7% to 18% of India's 2019 GDP from 2016 to 2050. This shift to a low-carbon economy will necessitate a comprehensive overhaul of India's economic investment and development priorities.

Government-backed climate funds can be pivotal in addressing these challenges and helping India achieve its net-zero target by 2050.

342. Providing Financial Support to Social Media Bloggers for Promoting Climate Action

In today's digital age, social media bloggers wield significant influence, with thousands or even millions of followers. Their ability to create impactful content can be a powerful tool for climate action by raising awareness of climate change and promoting sustainability.

Recognizing this potential, the Uttar Pradesh government has introduced a policy to support social media influencers in promoting government initiatives. By providing financial assistance, the government can enable bloggers to create climate-focused videos that inspire action, share solutions, and engage audiences.

Support could include grants for video production, sponsorship for climate content, training programs, access to climate experts, and social media advertising. This collaboration can drive climate awareness, encourage sustainable lifestyles, and foster community action.

343. Government to Promote Biogas Adoption for Sustainable Cooking and Energy in Rural Areas

The Ministry of New and Renewable Energy

(MNRE) launched the National Bioenergy Programme in November 2022, focusing on promoting biogas in rural areas. The program aims to establish small and medium-sized biogas plants to provide clean cooking fuel, generate power, and offer thermal energy, helping to reduce greenhouse gas emissions and improve sanitation.

With India's large livestock population, biogas has significant potential to meet rural energy needs and reduce dependence on traditional fuels like LPG. The program also helps create organic bio-manure, benefiting farmers by reducing chemical fertilizer use. Additionally, biogas can be used for cooking, lighting, and even as a green fuel for vehicles.

By increasing awareness and support for adopting biogas plants, the government can empower rural communities, promote sustainable energy solutions, and mitigate climate change.

344. Encouraging Green AI for Sustainable Solutions

Governments play a crucial role in encouraging the adoption of green AI for sustainable solutions to combat climate change. By implementing policies, incentives, and regulations, they can support the development of energy-efficient AI technologies aligned with environmental goals. This includes funding research, offering tax incentives for sustainable AI practices, and setting standards for energy-efficient data centers and AI systems.

Governments can also foster public-private partnerships to scale up green AI applications, such as optimizing renewable energy grids, enhancing climate modelling, and improving disaster prediction and response. Integrating green AI into national climate action plans helps reduce carbon emissions, enhance resource efficiency, and support the transition to a low-carbon economy.

Raising awareness and providing education and training ensures that industries and communities can harness the benefits of green AI. Through proactive leadership and collaboration, governments can position green AI as a cornerstone of sustainable development, helping to secure a greener, more resilient future.

International Level

"Climate change is a global challenge, and we need global action to address it."

Dr. P. V. Shukla, Environmental Scientist

Due to its global nature, addressing climate change requires coordinated international efforts. By working together, nations can pool resources, share knowledge, and implement more effective solutions than individual efforts. International collaboration is essential to mitigate climate change and ensure a sustainable future for all.

345. Enhancing Global Climate Action: Strengthening the Paris Agreement

Strengthening the Paris Agreement requires countries to enhance their Nationally Determined Contributions (NDCs) to keep global temperatures below 1.5°C. This includes setting more ambitious targets for emissions reductions, expanding renewable energy use, and improving climate resilience.

A Global Stocktake Mechanism should be implemented every five years. It allows nations to assess their progress and update their commitments based on the latest scientific findings to achieve the agreement's objectives.

346. Advancing Global Carbon Pricing for Emission Reduction

Implementing carbon taxes and emissions trading systems (ETS) can advance global carbon pricing. A global carbon tax would create financial incentives for businesses to reduce their carbon footprint by making polluting activities more expensive.

Simultaneously, an emissions trading system would allow countries or companies to trade carbon credits, enabling lower-emission nations to sell credits to higher-emitting ones, promoting cost-effective emissions reductions and encouraging clean energy innovation.

347. Strengthening International Green Finance for Climate Action

International green finance is crucial in addressing climate change by supporting the transition to a low-carbon economy. The Green Climate Fund (GCF) should receive increased funding from developed nations to help developing countries adopt clean energy solutions, adapt to climate impacts, and implement sustainable practices.

Countries and companies can issue climate bonds to raise funds for large-scale green infrastructure projects, such as renewable energy installations, sustainable agriculture, and climate-resilient infrastructure, ensuring long-term environmental sustainability.

348. Promoting Global Technology Transfer for Sustainable Development

Technology transfer is vital for promoting global sustainability and reducing emissions. Wealthier nations should support the transfer of clean energy technologies to developing countries, helping them adopt sustainable energy systems and reduce their environmental impact.

Furthermore, international collaboration on the research, development, and deployment of renewable energy technologies like solar, wind, and battery storage can accelerate the global transition to clean energy, ensuring a more sustainable future for all nations.

349. Global Reforestation and Biodiversity Conservation for Climate Action

Global reforestation and conservation initiatives are crucial for combating climate change and preserving biodiversity. Large-scale reforestation and afforestation efforts, driven by international cooperation, can significantly reduce CO_2 levels in the atmosphere and restore vital ecosystems. Protecting carbon sinks like forests, wetlands, and mangroves is essential for long-term environmental health.

Additionally, biodiversity conservation through global agreements ensures the protection of ecosystems that play a key role in carbon sequestration, helping maintain ecological balance and enhance climate resilience.

350. Supporting Adaptation and Resilience in Vulnerable Nations

Adaptation and resilience support are essential for helping vulnerable nations cope with the effects of climate change. Climate adaptation funds from developed nations can aid small island developing states (SIDS) and low-income countries build resilience against extreme weather events, sea-level rise, and droughts.

Additionally, international cooperation in disaster risk reduction (DRR) can strengthen infrastructure and improve risk management, reducing the impacts of climate-related events and enhancing global climate resilience.

351. Fostering Global Research and Data Sharing for Climate Action

International collaboration on research and data sharing is crucial for addressing climate change. Global climate research networks, such as those led by the Intergovernmental Panel on Climate Change (IPCC), can foster collaboration on understanding climate impacts, developing mitigation technologies, and crafting adaptation strategies.

Additionally, establishing data-sharing platforms allows countries to track emissions, monitor climate progress, and evaluate the effectiveness of policies, ensuring informed decision-making and coordinated global action.

352. Accelerating the Transition to Clean Energy by Phasing Out Fossil Fuel Subsidies

Phasing out fossil fuel subsidies is key to accelerating the transition to clean energy. Countries should collaborate to eliminate these subsidies, which would make renewable energy sources more affordable and encourage the widespread adoption of clean technologies

Additionally, global financial institutions should be encouraged to divest from fossil fuel industries and redirect investments towards sustainable, low-carbon projects, supporting the transition to a greener economy.

353. Building Climate Resilience Through International Adaptation Plans

International climate adaptation plans should focus on building climate-resilient infrastructure, especially in vulnerable regions. This involves sustainable urban planning, agriculture, and efficient water management systems to help communities adapt to changing climate conditions.

Additionally, joint initiatives to protect coastal areas from rising sea levels, such as restoring mangroves and coral reefs, can safeguard vital infrastructure and protect vulnerable populations.

354. Advancing a Circular Economy for Sustainability

Promoting a circular economy involves reducing waste and enhancing recycling through collaboration between governments, industries, and businesses. Reusing, recycling, and remanufacturing products can minimize emissions linked to production and disposal.

Additionally, setting international standards for sustainable product design, focusing on durability, recyclability, and lower environmental impact, can further advance the transition to a more sustainable economy.

355. Global Cooperation on Reducing Methane Emissions

International agreements on methane emissions are essential to address this potent greenhouse gas. Global cooperation is needed to reduce methane emissions from key sources like agriculture, landfills, and the fossil fuel industry, as it offers significant near-term climate benefits.

356. Climate-Resilient Agriculture

Climate-resilient agriculture can be achieved through global collaboration to promote sustainable farming practices. Techniques like regenerative farming, agroforestry, and precision agriculture help reduce emissions, enhance food security, and improve resilience to climate change impacts.

357. Sustainable Transportation

Sustainable transportation can be promoted globally by encouraging the adoption of electric vehicles (EVs) through international standards and incentives. Developing and implementing sustainable fuels and technologies for shipping and aviation will reduce emissions in these high-impact sectors.

358. Education and Awareness

Promoting education and awareness on climate change through global climate literacy campaigns is essential to drive behavioural change. Additionally, engaging youth and indigenous communities in international climate decision-making processes ensures diverse perspectives and empowers these groups to contribute to sustainable solutions.

359. Corporate Accountability

Corporate accountability is key to addressing climate change. Implementing a global carbon disclosure standard would require companies to report their emissions transparently.

Encouraging multinational corporations to adopt sustainable supply chain practices can significantly reduce environmental impacts and promote responsible business operations.

360. Ocean Conservation

Ocean conservation is crucial for climate action. Strengthening international agreements like the High Seas Treaty can better protect marine ecosystems and biodiversity. Additionally, promoting blue carbon initiatives, which focus on conserving coastal ecosystems such as mangroves, seagrasses, and salt marshes, can significantly enhance carbon sequestration and help mitigate climate change.

361. Climate Justice and Equity

Climate justice and equity are essential for addressing the global impacts of climate change. Policies must prioritize the needs of vulnerable nations and communities, ensuring they receive support in adapting to climate challenges. Additionally, the Loss and Damage Fund should be operationalized to compensate countries disproportionately affected by climate change, helping them recover and build resilience.

362. International Monitoring and Accountability

International monitoring and accountability are crucial for ensuring that climate commitments are met. Transparent reporting systems should be established to monitor and verify emissions reductions, ensuring countries follow through on their climate pledges. Additionally, peer review mechanisms can be implemented to encourage countries to assess and improve their climate policies, ensuring accountability and progress toward global climate goals.

363. Global Behavioural Change Campaigns

Global behavioural change campaigns are essential to drive widespread climate action. Promoting sustainable consumption through international campaigns can help reduce overconsumption and encourage eco-friendly lifestyles. Additionally, encouraging a global shift toward plant-based diets can significantly reduce emissions from livestock production, contributing to a lower-carbon future. These efforts help individuals and communities make conscious choices that align with sustainability goals.

364. Geoengineering Research and Governance

Geoengineering research and governance are crucial for addressing climate change while minimizing potential risks. Developing an international governance framework for geoengineering technologies, such as solar radiation management, ensures these methods are applied responsibly and safely. Ethical considerations should guide research with transparency and global cooperation to ensure that geoengineering efforts do not have unintended negative consequences on the environment or vulnerable populations.

365. Strengthening International Institutions

Strengthening international institutions is vital for coordinated global climate action. Empowering the United Nations Framework Convention on Climate Change (UNFCCC) can enhance its capacity to lead and facilitate global efforts in addressing climate change. Additionally, establishing a Global Climate Council would provide a dedicated body to oversee, enforce, and hold nations accountable for their climate commitments, ensuring a more structured and effective global response.

Through these international solutions, countries can collaborate to combat climate change, reduce emissions, protect vulnerable populations, and build a sustainable future for the planet. Collaboration and commitment at all levels—governments, industries, and citizens—are essential for the success of these efforts.

References

Chapter I: Overview
Introduction:
NATIONAL ACCOUNTS STATISTICS 2018
- https://www.mospi.gov.in/sites/default/files/reports_and_publication/statistical_publication/National_Accounts/NAS18/Highlights.pdf
- Per capita national income across India from financial year 2015 to 2022, with estimates until 2024
- https://www.statista.com/statistics/802122/india-net-national-income-per-capita/
- Average salary in India (2024 data)
- https://www.timedoctor.com/blog/average-salary-in-india/
- National Broadcast Policy: TRAI talks connecting 100 million 'TV Dark' homes
- https://www.exchange4media.com/media-tv-news/43-of-indian-homes-dont-have-tv-sets-trai-133582.html
- India Washing Machine Market: Industry Analysis and Forecast (2024-2030) by Type, Capacity, End Use and Region
- https://www.maximizemarketresearch.com/market-report/india-washing-machine-market/127185/
- Greenhouse Gases: Key Contributors to Climate Change
- Climate Change: Atmospheric Carbon Dioxide BY REBECCA LINDSEY
- https://www.climate.gov/news-features/understanding-climate/climate-change-atmospheric-carbon-dioxide
- India Third Biennial Update Report to the United Nations Framework Convention on Climate Change: Ministry of Environment, Forest and Climate Change, Government of India, 2021

- **Chapter II: Causes**
- Special Report: Global Warming of 1.5 ºC
- https://www.ipcc.ch/sr15/chapter/chapter-1/
- Global warming of 1.5°C by © 2019 Intergovernmental Panel on Climate Change.
- https://www.ipcc.ch/site/assets/uploads/sites/2/2019/06/SR15_Full_Report_High_Res.pdf
- INFOGRAPHIC: India's energy-related CO2 emissions by ET Energy World

- https://energy.economictimes.indiatimes.com/news/coal/infographic-indias-energy-related-co2-emissions/72277641
- India Third Biennial Update Report to the United Nations Framework Convention on Climate Change: Ministry of Environment, Forest and Climate Change, Government of India, 2021
- Govt. of India Ministry of Power: Central Electricity Authority
- http://cea.nic.in/reports/monthly/installedcapacity/2024
- Roadmap for Access to Clean Cooking Energy in India by Sasmita Patnaik, Saurabh Tripathi, and Abhishek Jain
- https://niti.gov.in/sites/default/files/2019-11/CEEW-Roadmap_for_Access_to_Clean_Cooking_Energy_in_India-Report.pdf
- India 2020 Energy Policy Review
- https://niti.gov.in/sites/default/files/2020-01/IEA-India%202020-In-depth-EnergyPolicy_0.pdf
- Density of Population in India 1901-2011
- https://www.medindia.net/health_statistics/general/populationdensity.asp
- World Population by Country 2024 (Live)
- https://worldpopulationreview.com/
- Publication: What a Waste 2.0: A Global Snapshot of Solid Waste Management to 2050 by Kaza, Silpa Yao, Lisa C. Bhada-Tata, Perinaz Van Woerden, Frank
- https://openknowledge.worldbank.org/handle/10986/30317
- By the Numbers: New Emissions Data Quantify India's Climate Challenge By Subrata Chakrabarty
- https://www.wri.org/blog/2018/08/numbers-new-emissions-data-quantify-indias-climate-challenge
- State of Forest Report 2021
- https://fsi.nic.in/forest-report-2021
- Ministry of Environment, Forest and Climate Change, Govt. of India
- http://moef.gov.in/
- The Causes of Climate Change
- https://climate.nasa.gov/causes/

- **Chapter III: Effects**
- 2024 State of Climate Services
- https://public.wmo.int
- Record high temperatures in several parts of India; 48 deg C in Delhi
- https://economictimes.indiatimes.com/news/politics-and-nation/delhi-records-all-time-high-of-48-deg-c/articleshow/69728668.cms?from=mdr
- Humanitarian Snapshot Report 17 : Drought and heatwave as on 9th May 2019

- https://reliefweb.int/report/india/humanitarian-snapshot-report-17-drought-and-heatwave-9th-may-2019
- Drought by World Health Organization
- https://www.who.int/health-topics/drought#tab=tab_1
- India Heat Wave, Soaring Up to 123 Degrees, Has Killed at Least 36 By Mujib Mashal
- https://www.nytimes.com/2019/06/13/world/asia/india-heat-wave-deaths.html
- World Health Organisation
- https://www.who.int/healthtopics/heatwaves#tab=tab_1
- Cyclones & their Impact in India
- https://ncrmp.gov.in/cyclones-their-impact-in-india/
- Total loss due to Kerala floods is Rs 40,000 crore, says Minister EP Jayarajan
- https://www.thenewsminute.com/article/total-loss-due-kerala-floods-rs-40000-crore-says-minister-ep-jayarajan-88261
- Lakes formed from melting glaciers are making the Himalayas less stable By Owen King
- https://qz.com/india/1790633/meltwater-lakes-in-himalayas-are-accelerating-glacier-shrinkage/
- Himalayas: The climate change fight of the century by Bibek Bhattacharya
- https://www.livemint.com/news/india/himalayas-the-climate-change-fight-of-the-century-1550092989514.html
- India Witnessing average sea level rise of 1.7mm/year by Vishwa Mohan
- https://timesofindia.indiatimes.com/india/india-witnessing-average-sea-level-rise-of-1-7mm/year/articleshow/72134279.cms
- Sea-level rise could put 300 million people at risk by 2050 By Urmi Goswami
- https://economictimes.indiatimes.com/news/politics-and-nation/sea-level-rise-could-put-300-million-people-at-risk-by-2050/articleshow/71822608.cms
- Satellites help track ocean acidification in the Bay of Bengal by Sahana Ghosh
- https://india.mongabay.com/2019/12/satellites-help-track-ocean-acidification-in-the-bay-of-bengal/
- Dealing with the effects of ocean acidification on coral reefs in the Indian Ocean and Asia
- https://www.sciencedirect.com/science/article/pii/S2352485518306017
- Climate change impacts on ecosystem functions and services in India: An exploration of concepts and a state of knowledge synthesis

- https://www.researchgate.net/publication/334598699_Climate_change_impacts_on_ecosystem_functions_and_services_in_India_An_exploration_of_concepts_and_a_state_of_knowledge_synthesis
- Shifting Ecosystems
- https://scied.ucar.edu/longcontent/shifting-ecosystems
- Climate change and health: Why should India climate change and health: Why should India be concerned?
- http://www.bioline.org.br/pdf?oe09003
- India suffered 8% GDP loss in 2022 because of climate change: study
- Ground News - India suffered 8% GDP loss in 2022 because of climate change: study
- India may face 24.7% GDP loss by 2070 thanks to climate change: ADB report India may face 24.7% GDP loss by 2070 thanks to climate change: ADB report - The Economic Times
- Climate Change could affect India's GDP, warns report by Achintyarup Ray
- https://timesofindia.indiatimes.com/city/kolkata/climate-change-could-affect-indias-gdp-warns-report/articleshow/74543376.cms
- Impacts of Climate Change on Indian Agriculture
- https://dst.gov.in/sites/default/files/Report_DST_CC_Agriculture.pdf
- Impact of Climate Change on Agriculture
- https://pib.gov.in/PressReleaseIframePage.aspx?PRID=1909206
- With 23 Lakh Premature Pollution-Related Deaths Every Year, India Remains Most-Affected: Global Study by TWC India Edit Team
- https://weather.com/en-IN/india/pollution/news/2019-12-20-23-lakh-premature-pollution-deaths
- Global Burden of Disease Study 2017 by INSTITUTE FOR HEALTH METRICS AND EVALUATION
- http://www.healthdata.org/sites/default/files/files/policy_report/2019/GBD_2017_Booklet.pdf
- State of Global Air Report 2024
- https://www.stateofglobalair.org/sites/default/files/soga_2019_report.pdf

- **Chapter IV: Challenges**
- Increase in offences under 4 of 7 environment-related acts in 2022: NCRB
- https://www.downtoearth.org.in/environment/increase-in-offences-under-4-of-7-environment-related-acts-in-2022-ncrb-93222
- https://www.downtoearth.org.in/news/environment/environmental-crimes-india-may-take-upto-33-yrs-to-clear-case-backlog-says-report-71650

- DATA DRIVE: Huge pendency in green cases
- https://www.financialexpress.com/opinion/data-drive-huge-pendency-in-green-cases/1983035/
- Crime in India 2018 and 2023, report released by National Crime Records Bureau
- https://ncrb.gov.in/
- MINISTRY OF LAW AND JUSTICE
- http://nbaindia.org/uploaded/Biodiversityindia/Legal/31.%20Biological%20Diversity%20%20Act,%202002.pdf
- India Population (LIVE)
- https://www.worldometers.info/world-population/india-population/

- **Chapter V: Solution**
- India's Updated First Nationally Determined Contribution Under the Paris Agreement
- https://unfccc.int/sites/default/files/NDC/2022-08/India%20Updated%20First%20Nationally%20 20Determined%20Contrib.pdf
- Climate Action Tracker: https://climateactiontracker.org/countries/usa/
- **Individual Level:**
- Transforming India's Road Infrastructure
- https://static.pib.gov.in/WriteReadData/specificdocs/documents/2024/jul/doc2024726355201.pdf
- Live Green: 10 Simple Ways to Use Less Paper & Save the Planet's Forests
- https://www.thebetterindia.com/186651/india-steps-to-save-paper-recycle-live-green-india/
- Australians send 1.9 million tonnes of paper each year to landfill, most of which can be recycled...
- https://www.1millionwomen.com.au/blog/10-small-things-you-can-do-save-paper/
- Wasteful Packaging Is the Worst. Here's How to Push Back by Laura Newcomer
- https://greatist.com/live/reduced-packaging-how-to-reduce-waste-in-your-daily-life#1
- How Much Do Our Wardrobes Cost to the Environment? By María Fleischmann - World Ban
- E-waste and how to reduce it by Rebecca Adams
- A Complete Guide to Greening Your Laundry & Reducing your Carbon Footprint by THE BETTER HOME TEAM JULY 29, 2020
- https://www.thebetterindia.com/234048/complete-guide-reducing-carbon-footprint-green-eco-friendly-laundry-tips-the-better-home/

- Let's See How Many Trees You Can Save From Not Using Tissue Paper
- https://earthbuddies.net/trees-not-using-tissue/
- 21 crore LED bulbs distributed, to save Rs 11,000 crore: PM Narendra Modi
- https://economictimes.indiatimes.com/industry/energy/power/21-crore-led-bulbs-distributed-to-save-rs-11000-crore-pm-narendra-modi/articleshow/57024047.cms?from=mdr
- Upcycling: Save the planet for future generations by Andy Hayes, Michigan State University Extension
- What are bio enzyme natural cleaners? By Pooja Khanna Tyagi
- India has 22 cars per 1,000 individuals: Amitabh Kant by Muntazir Abbas
- Climate change: we can reclaim cities from the car without inconveniencing people by Richard Kingston and Ransford A. Acheampong
- What share of global CO_2 emissions come from aviation? By: Hannah Ritchie
- https://ourworldindata.org/global-aviation-emissions#
- Should we give up flying for the sake of the climate? By Jocelyn Timperley
- Number of operating vehicles in India FY 1951-2022
- https://www.statista.com/statistics/664729/total-number-of-vehicles-india/
- Irrigation in India: Status, challenges and options by RAJNI JAIN, PRABHAT KISHORE and DHIRENDRA KUMAR SINGH
- https://krishi.icar.gov.in/jspui/bitstream/123456789/34362/1/irrigation_rajni_preprint.pdf
- India needs to conserve waterbodies and value them By Mahreen Matto
- https://www.worldometers.info/world-population/world-population-by-year/
- Why family planning is key to climate adaptation by Bridget Kelly
- India is burning—here's how to stop it by CCAC secretariat , 11 November, 2019
- https://www.ccacoalition.org/en/news/india-burning%E2%80%93%E2%80%93here%E2%80%99s-how-stop-it
- Passenger vehicle makers miss fuel efficiency targets By Ashutosh R Shyam
- https://economictimes.indiatimes.com/industry/auto/auto-news/passenger-vehicle-makers-miss-fuel-efficiency-targets/articleshow/106617808.cms?from=mdr
- India GHG Program

- https://indiaghgp.org/
- Electric vehicle market in India expected to hit 63 lakh units per annum mark by 2027: IESA
- https://auto.economictimes.indiatimes.com/news/industry/electric-vehicle-market-in-india-expected-to-hit-63-lakh-units-per-annum-mark-by-2027-iesa
- Pradhan Mantri Ujjwala Yojana Fuels LPG Revolution Nearly 17 crore LPG consumers added in last 9 years
- https://pib.gov.in/PressReleasePage.aspx?PRID=1918364#
- LPG Revolution: 17 crore new connections double customer base in 9 years
- https://economictimes.indiatimes.com/industry/energy/oil-gas/lpg-revolution-17-crore-new-connections-double-customer-base-in-9-years/articleshow/99641712.cms?from=mdr
- Solar Cooking in India By Arushi Prakash
- https://www.ecoideaz.com/expert-corner/solar-cooking-in-india
- Toilets Built Under Swachh Bharat Mission
- https://pib.gov.in/PressReleaseIframePage.aspx?PRID=1797158
- Staying eco-friendly during the festivals
- https://careearthtrust.org/staying-eco-friendly-during-the-festivals
- Climate change is real ... ... turn off your computer!
- https://sustainability.tufts.edu/wp-content/uploads/Computer_brochures.pdf
- The Global Warming Survival Guide
- http://content.time.com/time/specials/2007/environment/article/0,28804,1602354_1603074_1603535,00.html
- Do-It-Yourself Home Energy Audits
- https://www.energy.gov/energysaver/home-energy-audits/do-it-yourself-home-energy-audits
- How Often Do You REALLY Need to Change Your HVAC Filter?
- https://ushomefilter.com/how-often-change-hvac-filter/
- Measure your green building performance with Arc in India by Apoorv Vij 22 Jun 2017
- https://gbci.org/measure-your-green-building-performance-arc-india
- How is Air Quality Measured?
- https://scijinks.gov/air-quality/
- Save the Planet, Clean Your Inbox By Terri Kafyeke
- https://en.reset.org/blog/save-planet-clean-your-inbox-12242015
- Digital Cleanup Day: Delete an email - save the planet, urge environmentalists By Euronews
- https://www.euronews.com/2020/04/21/digital-cleanup-day-declutter-your-devices-to-help-the-planet-urge-environmentalists

- Renewable Energy: Creating a Sustainable World
- https://www.investindia.gov.in/sector/renewable-energy
- 5 tech innovations that could save us from climate change
- https://www.weforum.org/agenda/2017/01/tech-innovations-save-us-from-climate-change/
- How to teach your children about climate change in small, but significant ways By Roopal Kewalya
- https://indianexpress.com/article/parenting/blog/how-to-teach-children-about-climate-change-in-small-but-significant-ways-5808263/
- Educating Girls is More Effective in the Climate Emergency than Many Green Technologies By Rapid Transition Alliance Staff
- https://www.resilience.org/stories/2020-02-24/educating-girls-is-more-effective-in-the-climate-emergency-than-many-green-technologies/
- Petition to declare climate change as an emergency in India
- https://www.change.org/p/harsh-vardhan-ministry-of-environment-forest-and-climate-change-petition-to-declare-climate-change-as-an-emergency-in-india
- Climate Warriors
- https://www.unicef.org/india/campaigns/climate-warriors
- India Focus: State Progress on Green Buildings By Anjali Jaiswal November 16, 2020
- https://www.nrdc.org/experts/anjali-jaiswal/india-focus-state-progress-green-buildings
- The ripple effect of public transit on carbon footprint reduction by Monalisha Thkur
- https://energy.economictimes.indiatimes.com/news/power/the-ripple-effect-of-public-transit-on-carbon-footprint-reduction/111775379#
- How banks and green finance are helping address climate change
- https://www.cfainstitute.org/insights/articles/how-banks-and-green-finance-are-helping-address-climate-change
- Climate Smart Agriculture: A Key to Sustainability
- https://loksabhadocs.nic.in/Refinput/New_Reference_Notes/English/31012022_115223_102120474.pdf
- How to Discuss Climate Change with Family & Friends
- https://www.wikihow.com/Talk-to-Your-Loved-Ones-About-Climate-Change
- How halting deforestation can help counter the climate crisis
- https://www.unep.org/news-and-stories/story/how-halting-deforestation-can-help-counter-climate-crisis
- Choice in electricity retail: Will consumers choose green power?

- https://www.orfonline.org/expert-speak/choice-in-electricity-retail-will-consumers-choose-green-power
- Why Birds Are Useful for the Environment
- https://walkintothewild.medium.com/why-birds-are-useful-for-the-environment-2f5115e755f4#
- What are the environmental benefits of keeping our beaches clean?
- https://baleforce.com/what-are-the-environmental-benefits-of-keeping-our-beaches-clean/#
- **Business Level:**
- Only 16% of global companies on track for 2050 net-zero goals: Accenture
- https://www.esgdive.com/news/16-percent-g2000-companies-on-track-net-zero-progress-2050-targets-accenture-report/732974/#
- Over 50% of firms committed to achieving net-zero emission target: PwC survey
- https://www.business-standard.com/industry/news/over-50-firms-committed-to-achieving-net-zero-emission-target-pwc-survey-124010900756_1.html
- Corporate Environmental Responsibility (CER) Circular by Ministry of Environment, Forest and Climate Change:
- http://environmentclearance.nic.in/writereaddata/public_display/circulars/OIEBZXVJ_CER%20OM%2001052018.pdf
- Pay 2% of capital investment for green clearance: Environment Ministry to Corporates https://economictimes.indiatimes.com/news/economy/policy/pay-2-of-capital-investment-for-green-clearance-environment-ministry-to-corporates/articleshow/64008830.cms?
- Forget CSR. It's time for CER: Corporate Environmental Responsibility
- http://unpacking.design/blog/corporate-environmental-responsibility/
- India's bold move towards emission control by By Heidi Vella on 18th March 2019
- https://www.power-technology.com/features/power-generation-in-india/
- Why Indian automobile industry needs to walk extra to clean up toxic emissions By Anumita Roychowdhury
- https://www.downtoearth.org.in/blog/environment/why-indian-automobile-industry-needs-to-walk-extra-to-clean-up-toxic-emissions-62067
- Corporate Social Responsibility in India by India Briefing **is produced** by Dezan Shira & Associates
- https://www.india-briefing.com/news/corporate-social-responsibility-india-5511.html/

- Corporate Social Responsibility (CSR) in India
- https://csrcfe.org/about-csr-in-india-public-policy/
- Net Zero Banking: Creating a long-term and sustainable financial services economy
- https://www.pwc.in/blogs/net-zero-banking-creating-a-long-term-and-sustainable-financial-services-economy.html#
- Over 20,000 Government schools in state get Rs 150-cr CSR funding boost by DHNS, Bengaluru, JAN 18 2020
- https://www.deccanherald.com/state/top-karnataka-stories/over-20000-government-schools-in-state-get-rs-150-cr-csr-funding-boost-795697.html
- How organizations can harness employee power for climate goals
- https://www.weforum.org/stories/2022/10/how-organizations-can-harness-employee-power-for-climate-goals/
- State Bank of India unit to raise Rs 2,000 crore for climate fund
- https://economictimes.indiatimes.com/industry/banking/finance/state-bank-of-india-unit-to-raise-rs-2000-crore-for-climate-fund/articleshow/115213505.cms?from=mdr
- How manufacturers can reduce carbon emissions
- https://oizom.com/how-manufacturers-can-reduce-carbon-emissions/
- 6 Unpredictable Ways of Reducing Waste Through Supply Chain Management
- https://www.threadinmotion.com/en/blog/6-unpredictable-ways-of-reducing-waste-through-supply-chain-management
- **Government Level:**
- Climate Change Education, Environmental Education in India, Climate Change, and its ambiguity in the National Education Policy by Sriranjini Raman December 21, 2020
- http://wjes.ca/climate-change-education-environmentaleducation-in-india-climate-change-and-itsambiguity-in-the-national-education-policy/
- India looking at $500 bn Investment in renewable energy generation by 2028 by Sudheer Singh
- https://energy.economictimes.indiatimes.com/news/renewable/india-looking-at-500-bn-investment-in-renewable-energy-generation-by-2028/69040344
- Creating a sustainable world
- https://www.investindia.gov.in/sector/renewable-energy
- Investing in innovation is key to net-zero emissions by
- Chris Henderson July 28, 2020
- https://www.c2es.org/2020/07/investing-in-innovation-is-key-to-net-zero-emissions/

- Time for India's net zero target by ET Edit in ET Editorials, December 13, 2020
- https://economictimes.indiatimes.com/blogs/et-editorials/time-for-indias-net-zero-target/
- India rejects $300 billion COP29 climate finance deal, calls it 'optical illusion'
- https://indianexpress.com/article/world/india-rejects-cop29-climate-finance-deal-calls-optical-illusion-9687012/
- A climate action plan: Here's a proposal India can take to London, and be perceived as a serious player by Raghuram Rajan February 22, 2021
- https://timesofindia.indiatimes.com/blogs/toi-edit-page/a-climate-action-plan-heres-a-proposal-india-can-take-to-london-and-be-perceived-as-a-serious-player/
- OPINION: Carbon tax and its impact on India by Divakar Vijayasarathy
- https://energy.economictimes.indiatimes.com/news/power/opinion-carbon-tax-and-its-impact-on-india/77373648
- Mapping India's Energy Subsidies 2020:
- https://www.iisd.org/system/files/publications/india-energy-transition-2020.pdf
- UN Chief Urges India To Kill Fossil Fuel Subsidies, End Coal Pledges After 2020
- https://science.thewire.in/environment/antonio-guterres-india-fossil-fuel-subsidies-coal-power-clean-energy-pandemic/
- PM – Surya Ghar: Muft Bijli Yojana by Ministry of New and Renewable Energy
- https://www.myscheme.gov.in/search/ministry/Ministry%20Of%20New%20and%20Renewable%20Energy
- Solar Panel Subsidy in India 2021
- https://www.loomsolar.com/blogs/collections/solar-panel-subsidy-in-india
- This Solar Home in UP Plans to 'Sell Electricity' to the Government Through Grid Tie Solar Installation by Sorav Shukla December 4, 2019
- https://solarclap.com/sell-electricity-to-the-government/#
- Govt launches PM E-DRIVE subsidy scheme for EVs with Rs 10,900 cr outlay
- https://www.business-standard.com/industry/auto/govt-launches-pm-e-drive-subsidy-scheme-for-evs-with-rs-10-900-cr-outlay-124100100833_1.html
- 353 Million LED Bulbs Distributed Under Government's UJALA Program by SOUMIK DUTTA JUL 08, 2019

- https://mercomindia.com/ujala-353-led-bulbs-sold/
- India launches LED bulbs for ⊚10 in rural areas by Utpal Bhaskar 19 Mar 2021
- https://www.livemint.com/news/india/india-launches-10-led-bulb-for-rural-areas-11616147268047.html
- Connecting India Like Never Before
- https://static.pib.gov.in/WriteReadData/specificdocs/documents/2024/jul/doc2024726355201.pdf
- Solar energy generation potential along national highways by Pragya Sharma & Tirumalachetty Harinarayana 04 April 2013
- https://link.springer.com/article/10.1186/2251-6832-4-16
- South Delhi Municipal Corporation to set up solar farms and drains to generate clean energy by Pras Singh Aug 21, 2018
- https://timesofindia.indiatimes.com/city/delhi/south-corpn-to-set-up-solar-farms-and-drains-to-generate-clean-energy/articleshow/65480177.cms
- CLIMATE FINANCE IN INDIA 2023
- 20231128_Climate-Finance-in-India2023.pdf
- 101 WAYS TO FIGHT CLIMATE CHANGE By Patrick Sisson, Megan Barber, and Alissa Walker
- 50 THINGS WE CAN DO ABOUT CLIMATE CHANGE AS INDIVIDUALS BY *Kira Simpson*
- https://www.mea.gov.in/articles-in-indian-media.htm?dtl/32018/Indias_Climate_Change_Policy_ Towards_a_Better_Future

My Green Journey...

- During a challenging period (March 2011), I met Mr. Ramesh Panda from Kalinga Ashram. After listening to his charity for orphans, it was then that I committed to serving people for the rest of my life. On 5th May 2011, I joined Repco Home Finance as a Branch Manager in Bhubaneswar. Alongside my wife, we began visiting orphanages and old age homes, spending meaningful time with children and the elderly, which brought us immense joy.

- In June 2015, during a discussion with Mr. Ramesh Panda at Kalinga Ashram, Berhampur, he shared that planting a tree is the most incredible charitable act and encouraged me to start a plantation initiative in my area. This advice was also supported by a Baba from Mahendragiri, who was present during the conversation.

- On 3rd July 2015, I launched a plantation drive in my village with the help of my village club members, naming the initiative "Green Parikud." Over time, this effort grew as more village members, school and college students, teachers, government officials, panchayat representatives, and Self Help Group (SHG) members joined.

- Gradually, this plantation activity spread to neighbouring villages and regions in Parikud, Puri, Bhubaneswar, and other states.

- In August 2016, I organized the first "Green Marathon" at Gopinath Dev Bidyapitha, Charichhak, Krushnaprasad, Puri, Odisha, to raise environmental awareness among school and college students, the local community, and block administration. A second Green Marathon followed in August 2017 at Trilochaneswar Bidyapitha, Tichini, Krushnaprasad, Puri, Odisha, continuing the push for environmental consciousness across various groups.

- The blessings of my plantation efforts were evident when I welcomed my first child on 16th September 2015. For his first birthday in 2016, I distributed 2,000 hybrid mango and lemon trees, worth Rs. 1,00,000, to my relatives, villagers, and neighbours who would have attended the celebration. We marked the occasion with a simple Satyanarayan Puja instead of a grand party, and those around me greatly appreciated the initiative.

- In the same year, my favourite school teacher, Simanchal Sir, advised me to allocate 15% of my monthly salary towards social work. I embraced this advice and made it a priority to support plantation and environmental awareness through this fund. This decision brought happiness and recognition in both my personal and professional life.

- In January 2018, I was honoured with the prestigious Dronacharya Award in my organization for outstanding performance. By April 2018, I had received top ratings, a significant increment, and a bonus. Fuelled by this success, I increased my contributions—allocating 20% of my monthly salary and 50% of my annual bonus to social causes, especially environmental efforts.
- Even today, I remain actively involved in plantation activities during weekends, holidays, and annual leave—wherever I am posted. The Green Parikud team continues to work year-round with the dedicated involvement of SHG women and like-minded environmental enthusiasts.
- During my tenure in Raipur, Chhattisgarh, I founded a group called "Climate Change Solution," the name of my book. This group brought together banking, services, and business professionals to champion environmental causes.
- Through the joint efforts of Green Parikud and Climate Change Solution, we have successfully planted 70,000 trees, 110,000 palm trees, and 100,000 neem seeds across Odisha and Chhattisgarh by the end of 2024—without any financial support from the government or NGOs.
- At an environmental awareness program at Ruknidevi Chilika Mahavidyalay, two journalists, Mr. Durga Swain and Mr. Kailash Pattnaik, encouraged me to write something for school and college students. Mr. Sakyasingha Mahapatra, founder of SakRobotix, also advised me to explore and write about climate change. I focus on climate change. This sparked my curiosity, and I researched the environmental crisis more thoroughly. I soon realized that climate change is humanity's greatest threat, and I decided to write a book on the subject.
- In January 2020, I began collecting books and articles on climate change. During the COVID-19 lockdown, I dedicated more time to reading and writing. Over the last five years, I spent at least 2-3 hours a day and entire weekends working on a book, reading 15–20 books and countless articles to deepen my understanding of the climate crisis.
- I firmly believe that trees are a form of God—they provide oxygen and absorb CO2. With their blessings, I have found true happiness, free from tension, despite the daily challenges in my personal and professional life. My relationships with family, friends, colleagues, and society remain strong, and those involved in my green journey are respected and admired.

I extend my heartfelt gratitude to my wonderful wife for her unwavering support throughout this journey and for lovingly caring for our two sons and family. I humbly ask her to continue standing by me as I carry this mission forward—for a greener, better world.

Black Eagle Books

www.blackeaglebooks.org
info@blackeaglebooks.org

Black Eagle Books, an independent publisher, was founded as a nonprofit organization in April, 2019. It is our mission to connect and engage the Indian diaspora and the world at large with the best of works of world literature published on a collaborative platform, with special emphasis on foregrounding Contemporary Classics and New Writing.